平面设计中的
数字版式设计

PINGMIAN SHEJI ZHONG DE
SHUZI BANSHI SHEJI

（美）罗杰·福塞特-唐 著
（Roger Fawcett-Tang）
廖坤 于钶 译

重庆大学出版社

图书在版编目（CIP）数据

平面设计中的数字版式设计 / （美）罗杰·福塞特–唐（Roger Fawcett-Tang）著；廖坤，于钶译. —重庆：重庆大学出版社，2016.11
（西学东渐·艺术设计理论译丛）
书名原文：Numbers in Graphic Design
ISBN 978-7-5624-9839-1

Ⅰ. ①平… Ⅱ. ①罗…②廖…③于… Ⅲ. ①数字—字体—版式—设计 Ⅳ. ①TS881②J292.13③J293

中国版本图书馆CIP数据核字（2016）第130718号

平面设计中的数字版式设计
PINGMIAN SHEJI ZHONG DE SHUZI BANSHI SHEJI
（美）罗杰·福塞特–唐 著
（Roger Fawcett-Tang）
廖 坤 于 钶 译

策划编辑：席远航
责任编辑：席远航　　书籍设计：胡 越
责任校对：张红梅　　责任印制：赵 晟
*
重庆大学出版社出版发行
出版人：易树平
社址：重庆市沙坪坝区大学城西路21号
邮编：401331
电话：（023）88617190　88617185（中小学）
传真：（023）88617186　88617166
网址：http：//www.cqup.com.cn
邮箱：fxk@cqup.com.cn（营销中心）
全国新华书店经销
重庆高迪彩色印刷有限公司印刷
*
开本：787mm × 1092mm　1/16　印张：20　字数：166千
2016年11月第1版　2016年11月第1次印刷
ISBN 978-7-5624-9839-1　定价：88.00元

0123456789——这10个数字对于设计师来说具有普遍的吸引力。它们具有多语种性、便于理解、形式多样、适应性强、而且能让人们过目不忘。因为这10个数字中的每一个都具有其牢固且独一无二的特性。

数字在平面设计中应用的重点在于平面设计师如何处理包含巨大信息量的数字的排序，如何从时间表到年度报告，从数据丰富的文档到极富创造力和有趣的排版实验中探索探索数字信息最抽象的概念。

本书分为8个章节来阐释数字的无处不在。添加（Addition）部分展示了图表显示和打印页面内的各式时间表，同时揭示了时间表是如何用来组织诸如插图、对象、交互式媒体和书面文字等可变数据的。数据（Data）部分则着眼于阐释各种各样的信息图案。尽管数字并不总是设计中的”主角”，但它们确实构成了所有插图设计系统的基础和结构要素。顺序（Order）部分以公交车、火车及电车网络时刻表附带各种事件列表及行程安排为特点，不仅突出日常实践中清晰的表格式排版，同时又突出了更富于表现力的解决方案。年表（Chronology）关注的则是时间：从日历、日志到钟表和各种屏幕上的解决方案，都用于图表与表达秒、分和时的更迭。抽象（Abstraction）揭示了各种进行数字表达的创意方向，以及在仍保持字符本质的前提下排版形式可达到的推进程度。地方性（Vernacalar）排版解决了日常生活中存在的数字显示形式。形式（Form）关注于设计中从正式到装饰的某一系列文本中的某一数字或一系列数字。增加（Multiplication）作为围绕建筑的引导标示与指示系统内部结构的导航，对打印页面上的超大形数字进行了说明，增加了结构及内容的规模感。最后，减少（Substraction）着眼于在维持可识别性的前提下数字的简化程度——以最简的数字来显示最丰富的信息。

最近有人问我最喜欢的数字是什么，我回答道：“这取决于字体；这10个数字我都非常喜欢。”在用清晰而有逻辑的方式来传达极度详细的信息方面，数字具有强大的力量——如处理复杂的时间表或年度报告。它们也能帮助你游走于建筑或书页之中。

01. 添加 007
01.1 时间轴 008
01.2 计算方法 038

02. 数据 043
02.1 量化 044
02.2 统计 066
02.3 报告与说明 082

03. 顺序 089
03.1 时刻表 090
03.2 列表 102

04. 年表 117
04.1 日历 118
04.2 日记簿 138
04.3 钟表 146

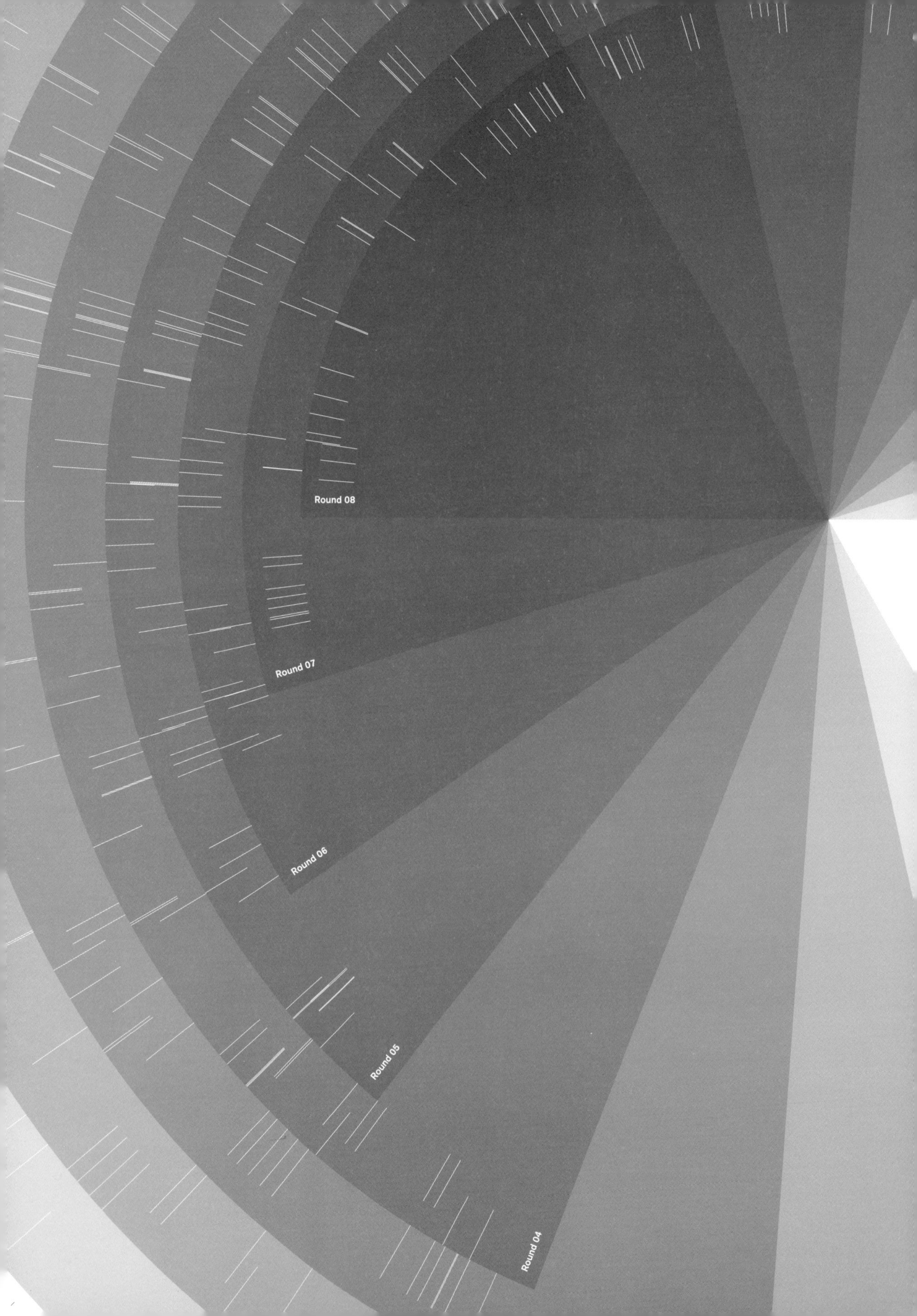
Round 08
Round 07
Round 06
Round 05
Round 04

01. 添加

01.1 时间轴
01.1.1 三维认知 008
01.1.2 交互 012
01.1.3 视窗 014
01.1.4 富有表现力的线条 016
01.1.5 笔触与线条 018
01.1.6 同轴韵律 020
01.1.7 光学噪声 022
01.1.8 图形弧线 024
01.1.9 网状图形 026
01.1.10 线性 028
01.1.11 数据尺度 032
01.1.12 尺度感 034
01.1.13 时效性 036

01.2 计算方法
01.2.1 时间间隔 038
01.2.2 时间流逝 040

01.1
时间轴

01.1.1
三维认知

展览会上，三维时间轴的延伸展示了“一级方程式赛车比赛”的历史。地板上，每个对应的年代都用大字标出，相关的赛车均按时间轴上所标注的位置进行放置。墙上是一些信息面板和手工艺品，如竞赛方案及历代一级方程式赛事的门票等。

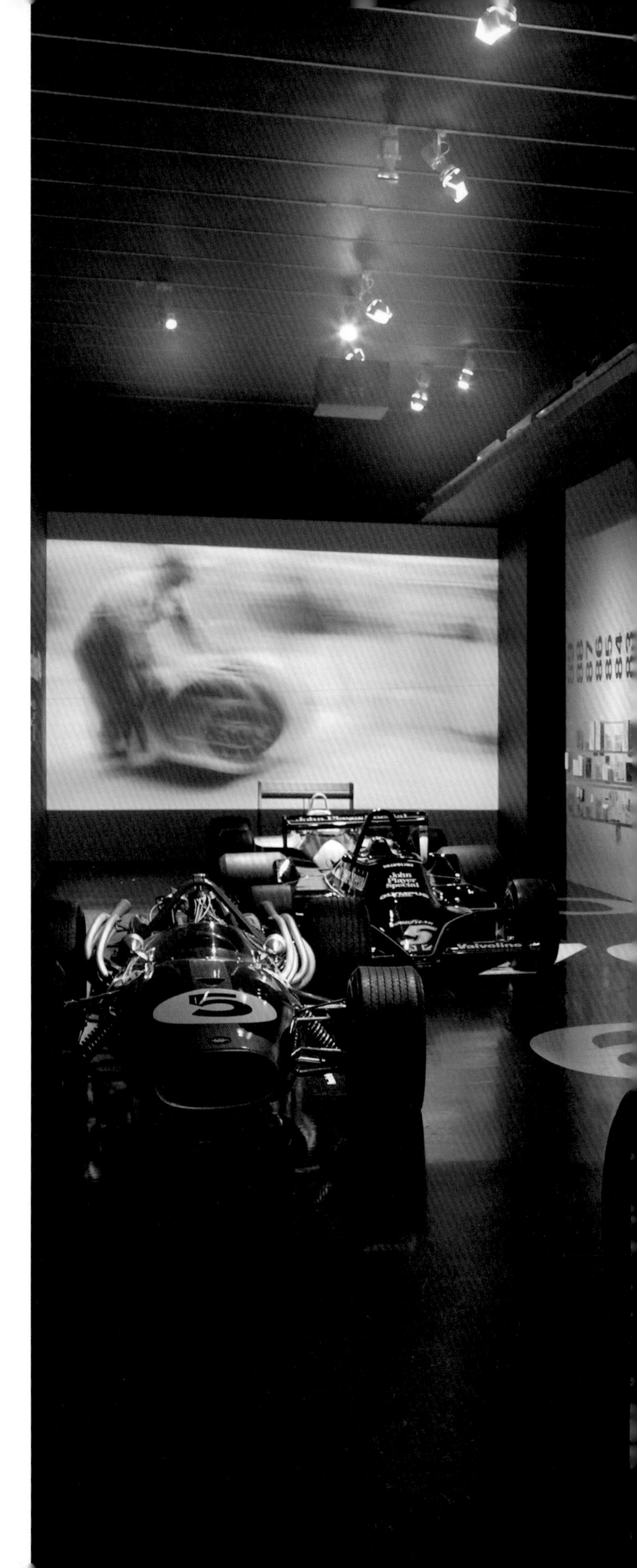

01

设计者：Studio Myerscough
作品名：*Formula One*
伦敦设计博物馆
巡回展览

73 72 71 70 69 68 67 66 65 64 63 62 61 60 59 58 57 56 55 54 53 52 51 50
First Grand Prix, 1950
24
24

01.1.1
三维认知

在“超级当代展”的画廊墙壁上，采用三条木制围栏来构成时间轴的结构，摆脱了用自粘性乙烯基材料印字这一传统方式。每条横木围栏边缘的凹槽内可以沿时间轴放入信息卡片、图片和手工艺品。顶部的横木围栏放置了标有主题的黑色大信息板，搭配着标有年代的红色小卡片。突破了时间轴常规和单纯的线性属性，较大的家具用品放置在木制围栏向下延伸出的更大的木制平台上。

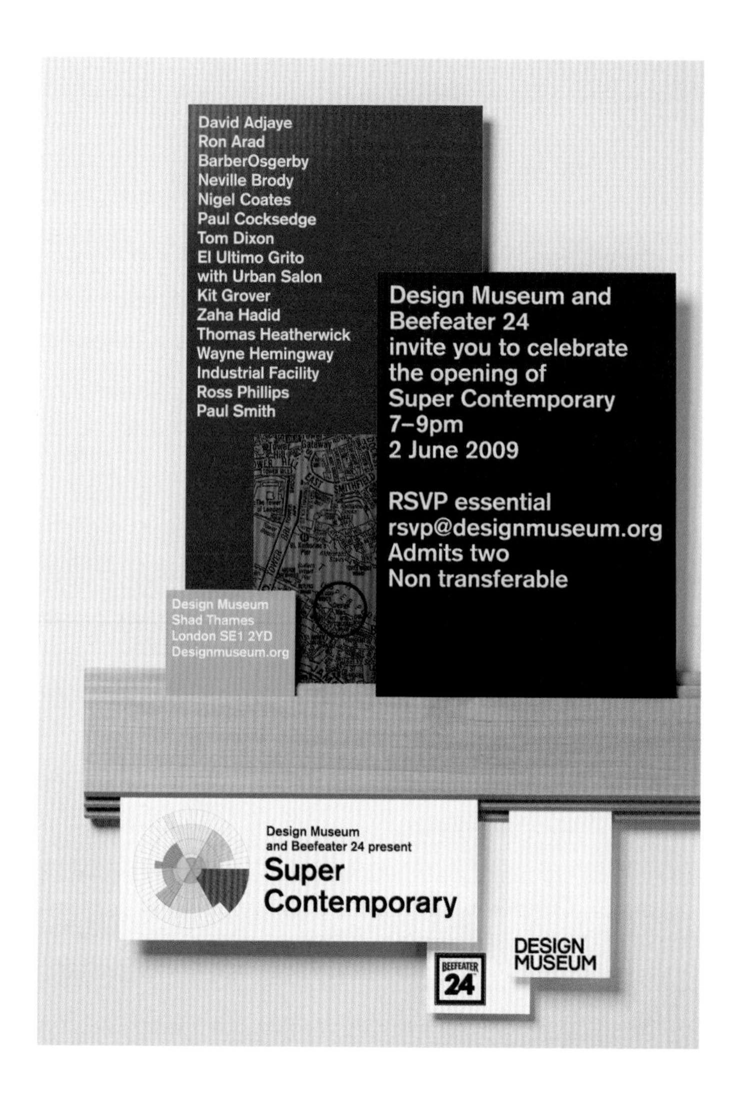

▸ 02
作品名：*Bibliotheque*
大英超级博物馆

Reviving familiar brands
2002
Multicultural melting pot
2003
Solid ground
2004
ECSTACITY
habitat

1962
Alternative approaches

A new consumer
1964
JAMES BOND BACK IN ACTION!
SEAN CONNERY
JAMES BOND 007
GOLDFINGER

01.1.2
交互

在“WerkStadt”对话框中，传统的展览时间轴面板与基于屏幕的交互式元素相结合。当屏幕沿着由玻璃板覆盖的信息板移动至适当内容部分时，就会触发相关视频的播放。这种基于屏幕的界面还会映射出静态展示图像的设计和布局。

01

设计者：L2M3 Kommunikationsdesign
作品名：*WerkStadt Dialog*
梅赛德斯-奔驰汽车，戴姆勒集团

r den
un?
tagsüber
1980

2,9
17
40
46
50
61
65
98
100
144
148

1083
1100
1200

01.1.3
视窗

“伟大时代的赞助人”这一设计中，由白色展示板做成的网格有助于使印刷于平面上的不同级别的信息系统化。看上去朴素的单色风格通过轻质的无衬线体的排印来实现，同时，重要历史人物的浅灰色签名也为这一风格起到了补充作用。

以各种手写工件为特色的“视窗”，使展览的极简设计显得更为生动，这些陈列盒延伸边框上的历史人物黑白肖像也颇具特色。由于肖像印制于边框的两侧，观展者只有站在一定的角度才能看到这些肖像，因此当观展者直视展示窗口时，并不会转移他们的视线。

▸ 01
设计者：L2M3 Kommunikationsdesign
作品名：*Great Moments of a Patron*
柏林国家图书馆

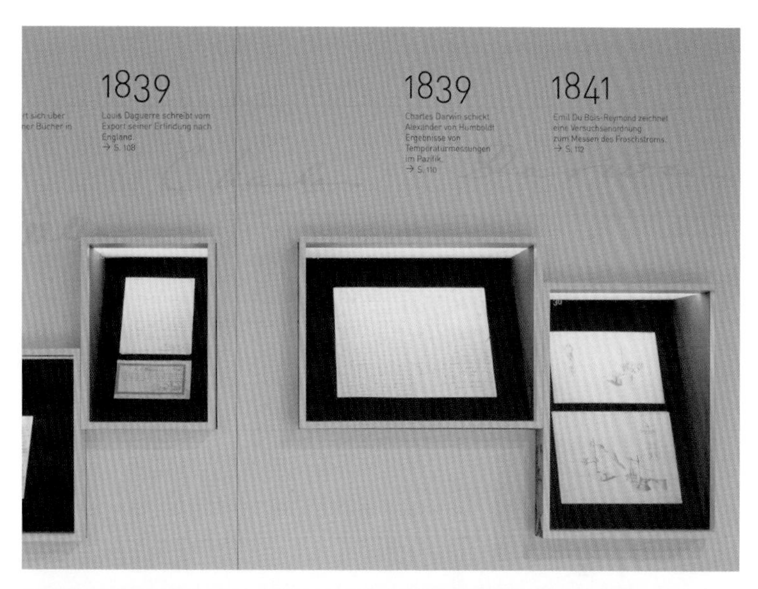
1839
Louis Daguerre schreibt vom Export seiner Erfindung nach England.
→ S. 108
1839
Charles Darwin schickt Alexander von Humboldt Ergebnisse von Temperaturmessungen im Pazifik.
→ S. 110
1841
Emil Du Bois-Reymond zeichnet eine Versuchsanordnung zum Messen des Froschstroms.
→ S. 112

Vor 100 Jahren schenkte Ludwig Darmstaedter der Königlichen Bibliothek seine Autographensammlung. Die Staatsbibliothek zu Berlin ehrt ihren großen Mäzen mit dieser Ausstellung.
60 Dokumente der berühmtesten Persönlichkeiten aus Wissenschaft, Kunst, Literatur und Geschichte sind Beispiele für die Universalität der über 250.000 Dokumente umfassenden Sammlung.
Machiavelli
Michelangelo
Kopernikus
1513
1519
1541
Wallenstein
Kepler
1621
1629
Galilei
Guericke
1634
1646
Descartes
Leeuwenhoek
Newton
Forster
Cook
Mendelssohn
Tilesius von Tiler
Napoléon
1806

01.1.4
富有表现力的线条

“K2运动时刻2010”系列限量版海报是为配合一家管理培训公司基于“比赛中的运动员”这一概念的研讨会课程而设计。每张海报都从经典赛事中提取数据元素，同时考虑速度、距离、时间等因素。胜利者被描绘成金色，而其他参赛者则被设计成灰色。无论何种情况下，抽象图形都阐明了事件的本质：两个同心圆代表着运动员在800米赛跑的过程中跑了两圈；波浪线表达游泳运动员身体周围的水流，等等。

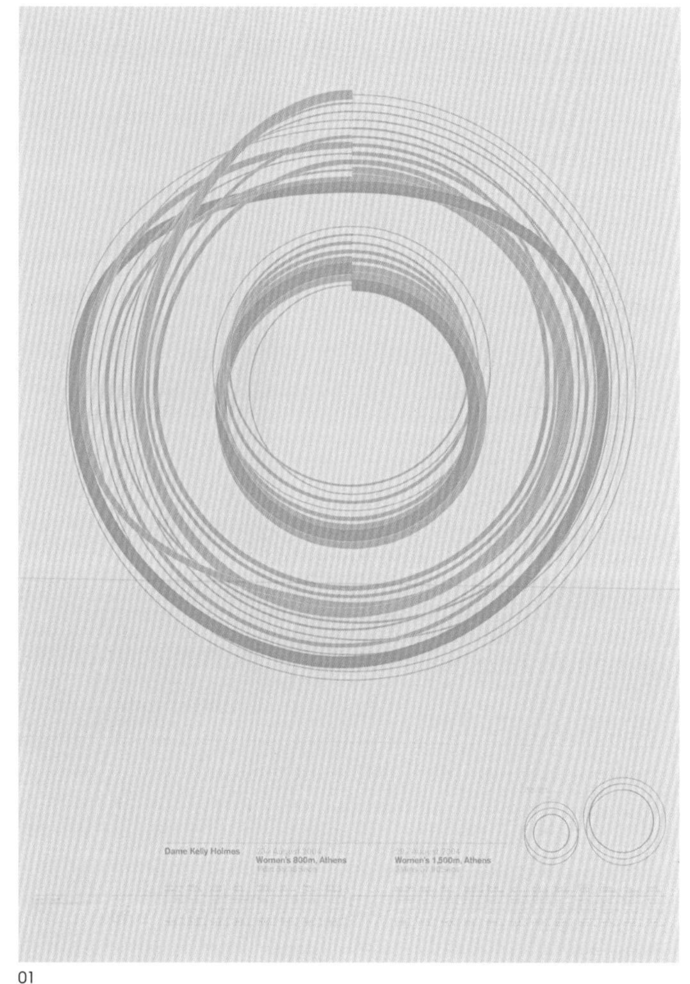

01

▸ 01

设计者：Accept and Proceed
作品名：*K2 Sporting Moments 2010*
Dame Kelly Holmes
女子800米比赛
2004年雅典奥林匹克运动会

▸ 02

设计者：Accept and Proceed
作品名：*K2 Sporting Moments 2008*
James Cracknell, Steve Redgrave,
Tim Foster and Matthew Pinsent
获得男子四人单桨无舵手艇项目金牌
2000年悉尼奥运会

▸ 03
▸ 04

设计者：Accept and Proceed
作品名：*K2 Sporting Moments 2010*
Mark Foster and Alexander Popov
男子50米自由泳
2003年巴塞罗那世界锦标赛

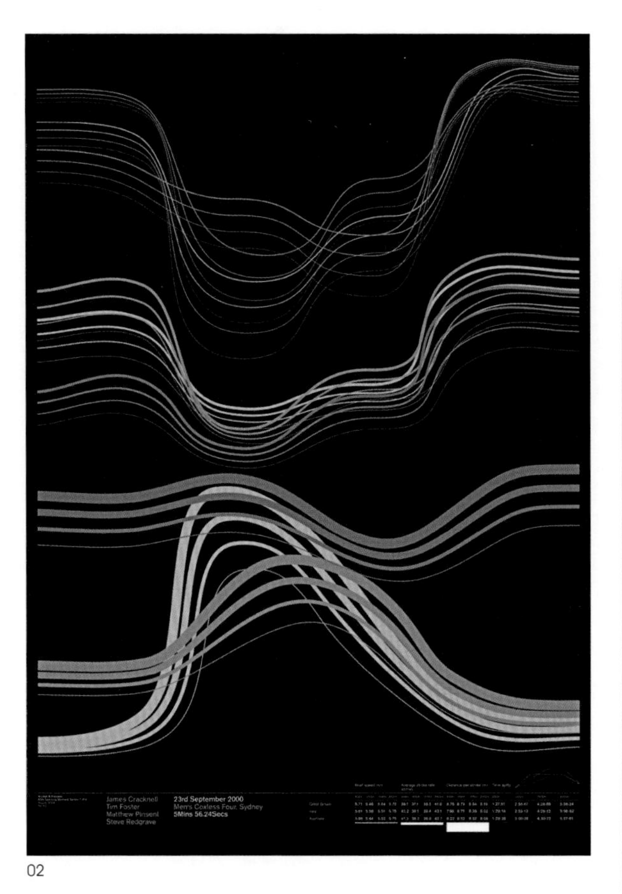
James Cracknell
Tim Foster
Matthew Pinsent
Steve Redgrave
23rd September 2000
Men's Coxless Four, Sydney
5Mins 56.24Secs

02

Alexander Popov
Mark Foster
Men's 50m Freestyle, Barcelona

03

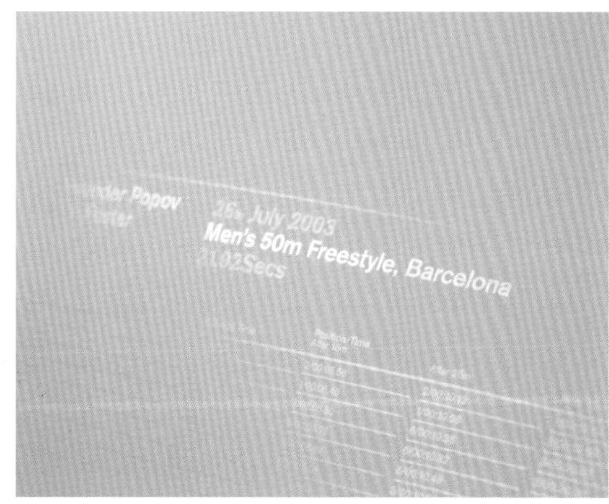
Men's 50m Freestyle, Barcelona

04

01.1.5
笔触与线条

这两张海报来自《比赛中的运动员》系列（见16页），它们描述的是2004年在圣地亚哥举行的美国高尔夫公开赛的第4天，泰格·伍兹对阵梅迪亚特。圆形的图表反映了每个高尔夫球手在那天的比赛中打出的杆数。第二张海报描述了2007年刘易斯·汉密尔顿在蒙特利尔大奖赛的首场胜利，这个图形反映出整个比赛期间每一位车手的位置移动情况。

01

02

▸ 01
▸ 02

设计者：Accept and Proceed
作品名：K2 Sporting Moments 2008
Tiger Woods
2008年美国高尔夫公开赛

▸ 03

设计者：Accept and Proceed
作品名：*K2 Sporting Moments 2008*
Lewis Hamilton
2007年蒙特利尔国际汽车大奖赛

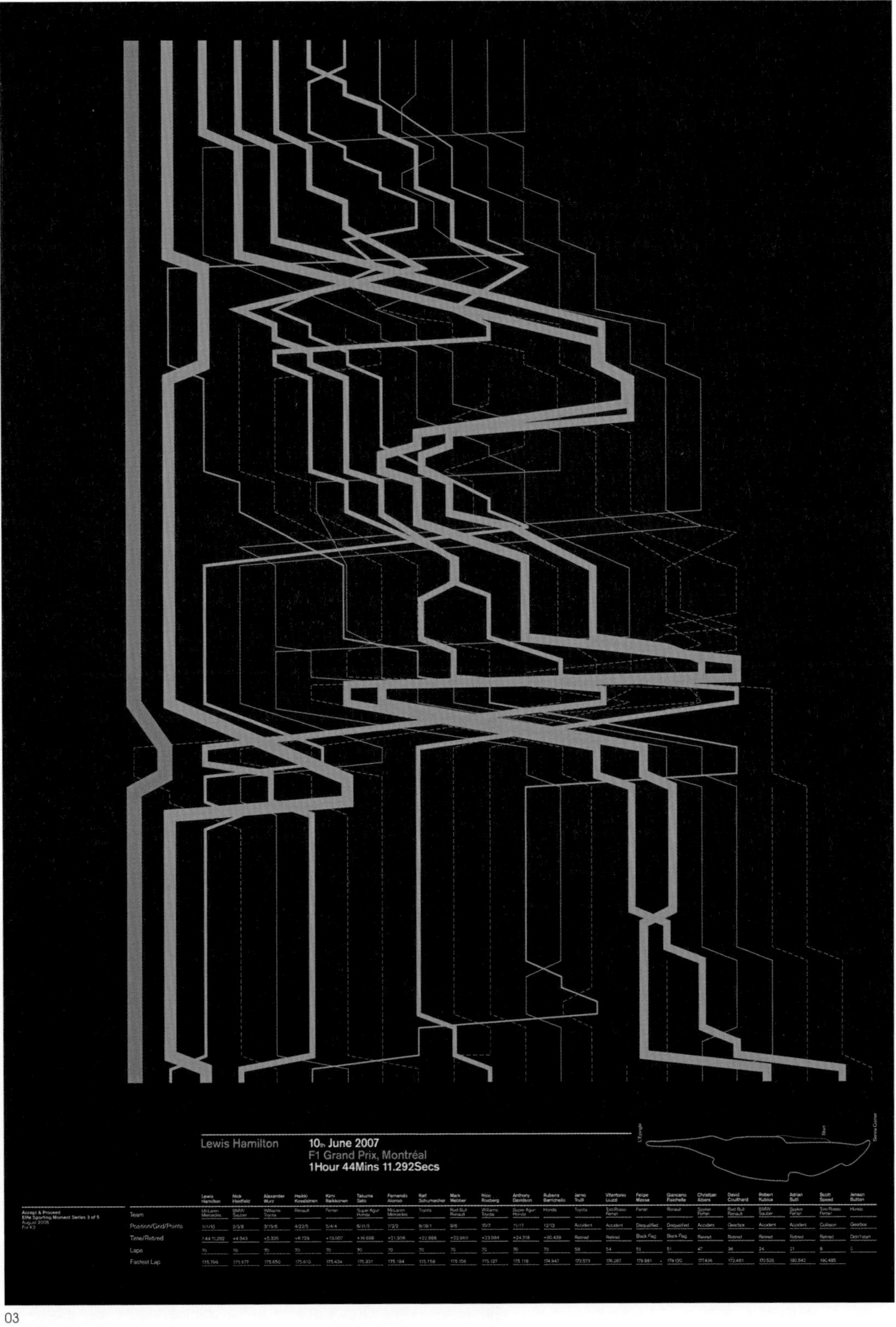
Lewis Hamilton
10th June 2007
F1 Grand Prix, Montréal
1Hour 44Mins 11.292Secs

03

01.1.6
同轴韵律

这张海报来自《比赛中的运动员》系列（见16–19页），其描述的是保罗·拉德克利夫在2005年伦敦马拉松赛中的胜利。重叠的同心圆反映的是运动员在伦敦市中心附近的26英里行程中的节奏和步伐。

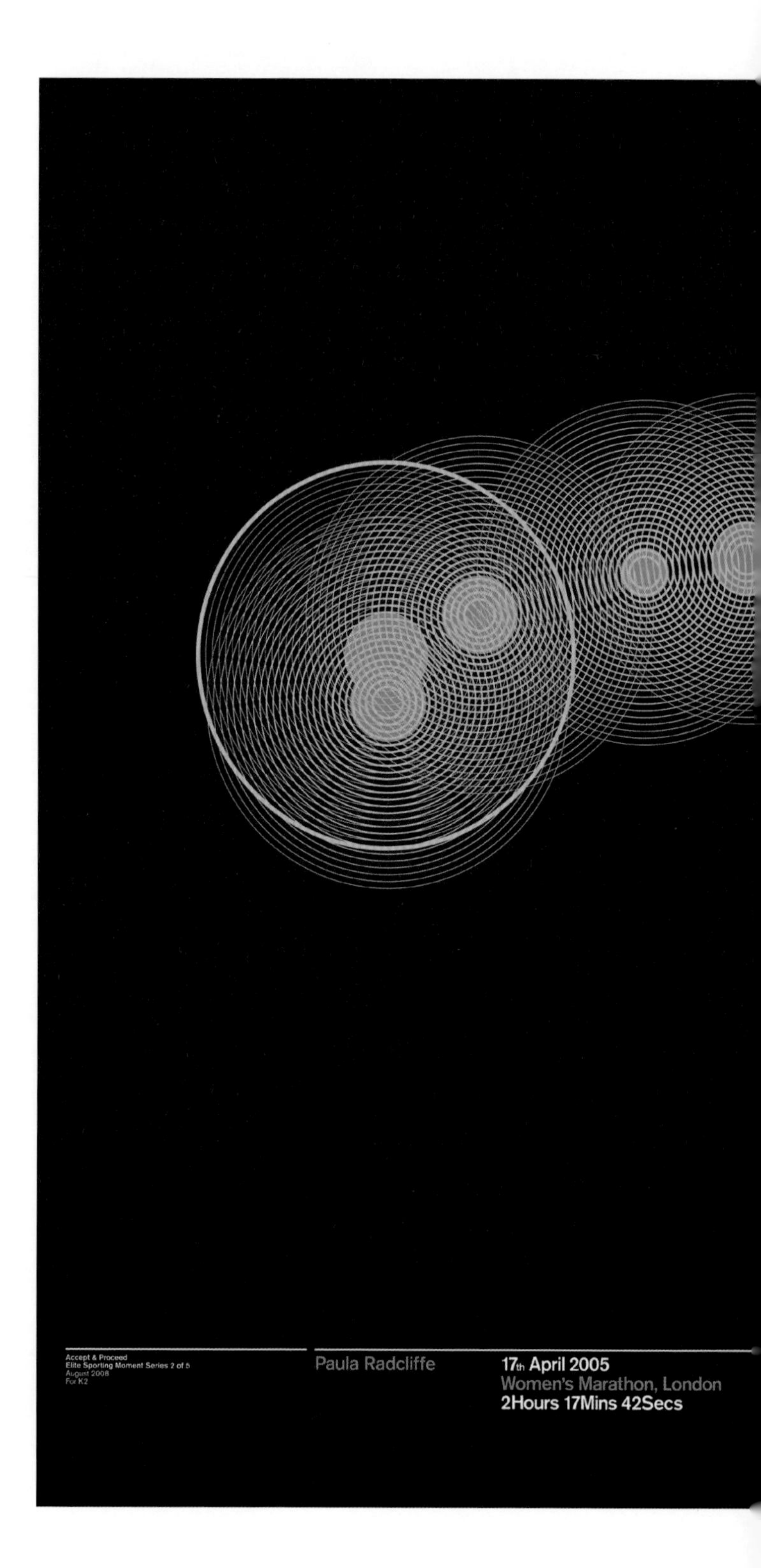

01

设计者：Accept and Proceed
作品名：*K2 Sporting Moments 2008*
Paula Radcliffe
2005年伦敦马拉松大赛

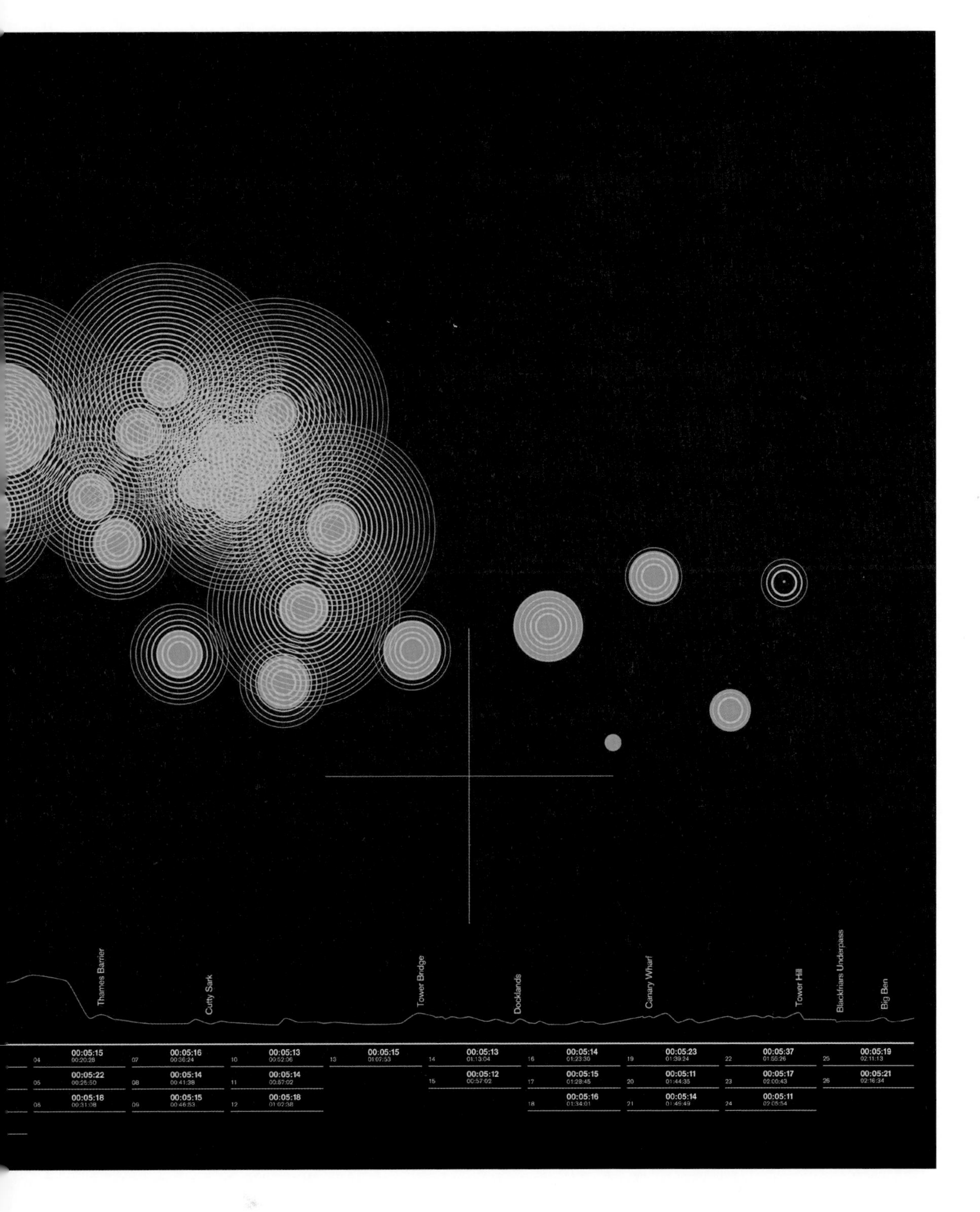
Thames Barrier
Cutty Sark
Tower Bridge
Docklands
Canary Wharf
Tower Hill
Blackfriars Underpass
Big Ben
04 00:05:15 00:20:28
05 00:05:22 00:25:50
06 00:05:18 00:31:08
07 00:05:16 00:36:24
08 00:05:14 00:41:38
09 00:05:15 00:46:53
10 00:05:13 00:52:06
11 00:05:14 00:57:02
12 00:05:18 01:02:38
13 00:05:15 01:07:53
14 00:05:13 01:13:04
15 00:05:12 00:57:02
16 00:05:14 01:23:30
17 00:05:15 01:28:45
18 00:05:16 01:34:01
19 00:05:23 01:39:24
20 00:05:11 01:44:35
21 00:05:14 01:49:49
22 00:05:37 01:55:26
23 00:05:17 02:00:43
24 00:05:11 02:05:54
25 00:05:19 02:11:13
26 00:05:21 02:16:34

01.1.7 光学噪声

两张一组的海报——《2007年的夜晚时长和2007年的日照时长》绘制出2007年间伦敦的所有白天时长。日界线显示为：一系列的同心圆代表白天，而水平线代表黑夜。当年的所有数据均用磷光油墨印在每张海报的底部。

01

02

▸ 01
▸ 02

设计者：Accept and Proceed
作品名：*Hours of Dark 2007*
A1纸丝网印刷

▸ 03

设计者：Accept and Proceed
作品名：*Hours of Light 2007*
A1纸丝网印刷

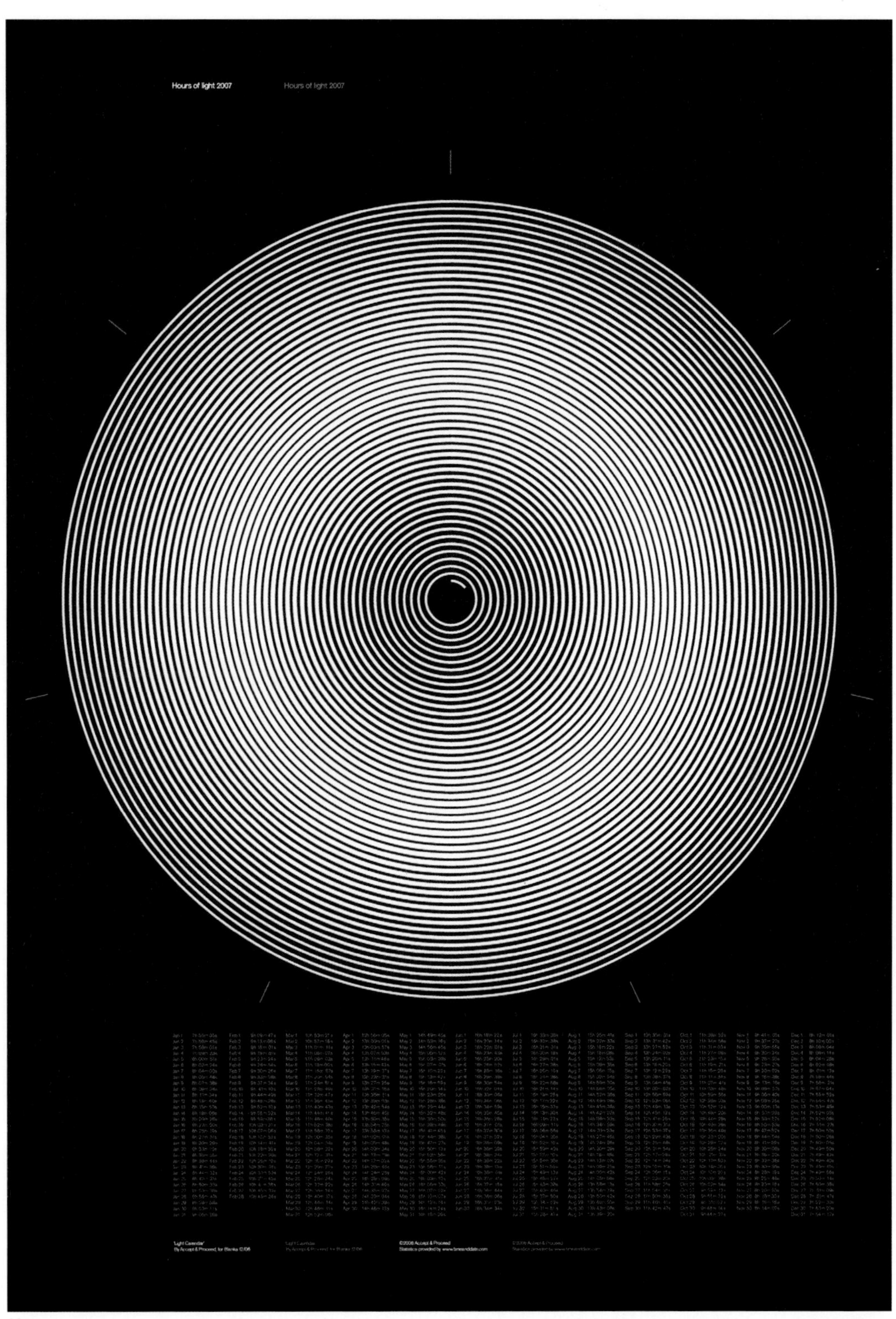
Hours of light 2007
'Light Calendar'
By Accept & Proceed, for Blanka 12/06
©2006 Accept & Proceed
Statistics provided by www.timeanddate.com

03

01.1.8
图形弧线

“Accept and Proceed”设计工作室继昼夜时长系列海报后（见22~23页），又创作了《2008年的日照时长》。“Image Now”工作室出品的A1海报描述了1974年穆罕默德·阿里与乔治·福尔曼这两位拳击手历史性战役的时间轴。每一条圆弧代表一个回合，同时标明出拳次数。

01

▸ 01

设计者：Accept and Proceed
作品名：Hours of Light 2008
A1纸丝网印刷

▸ 02

设计者：Image Now
穆罕默德·阿里
V.乔治·福尔曼
金沙萨，扎伊尔，1974年10月30日

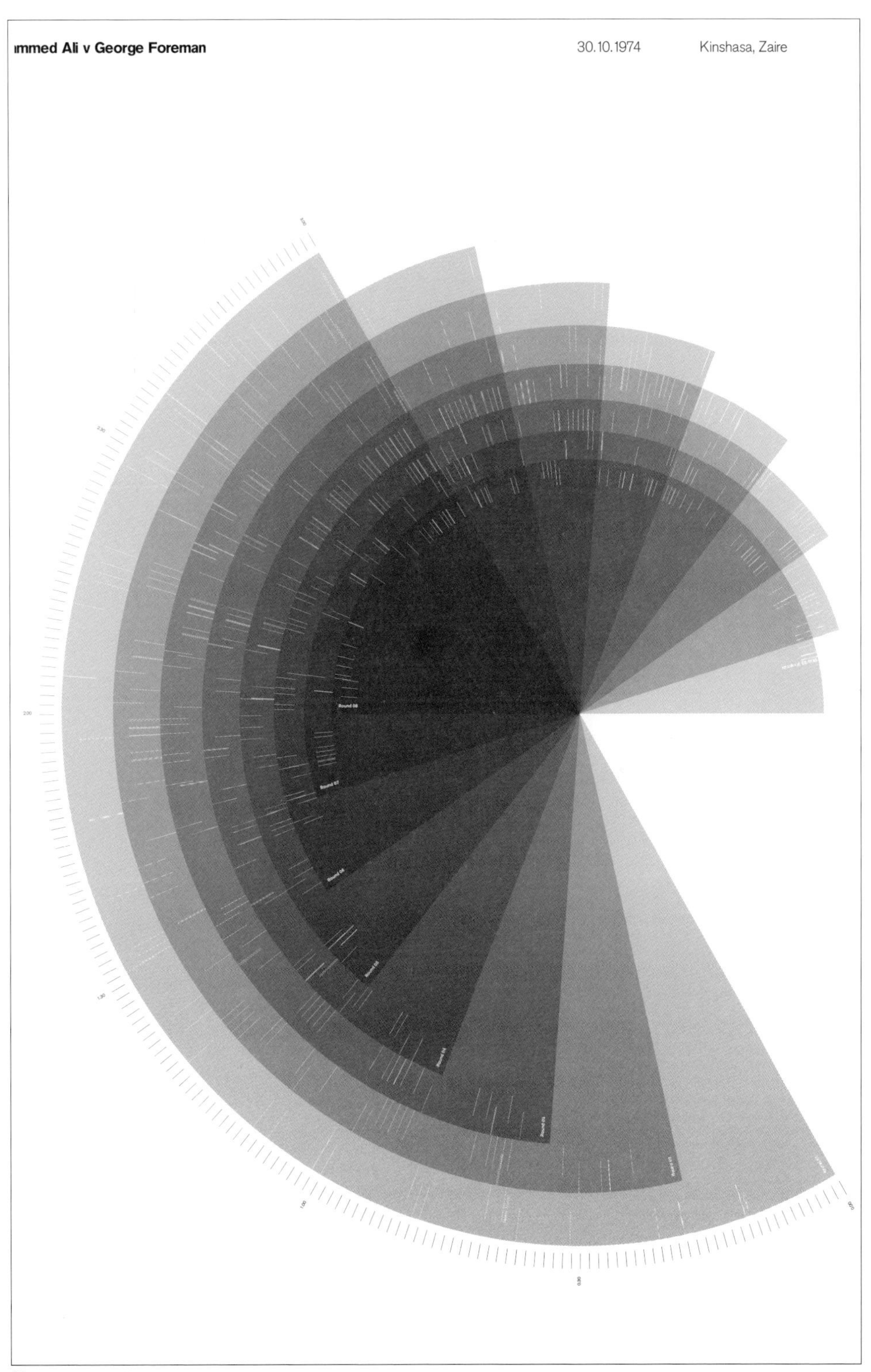

02

01.1.9
网状图形

图中的时间轴代表了从1998年乔斯特·格罗特斯（Joost Grootens）第一本书籍装帧设计开始，在超过十年时间中其书籍装帧设计的开展顺序。编号为1~100的小图标可对不同书籍进行识别，它们出现在乔斯特·格罗特斯的《我发誓我不使用任何艺术》一书中，用于指出哪本书是被提及的。一条简单的日界线穿过书本底部，而所有的书本信息排列在这些位置下侧。一系列的连接线显示出一个项目如何引出另一个项目，以及该时间周期内不同项目之间的关系。

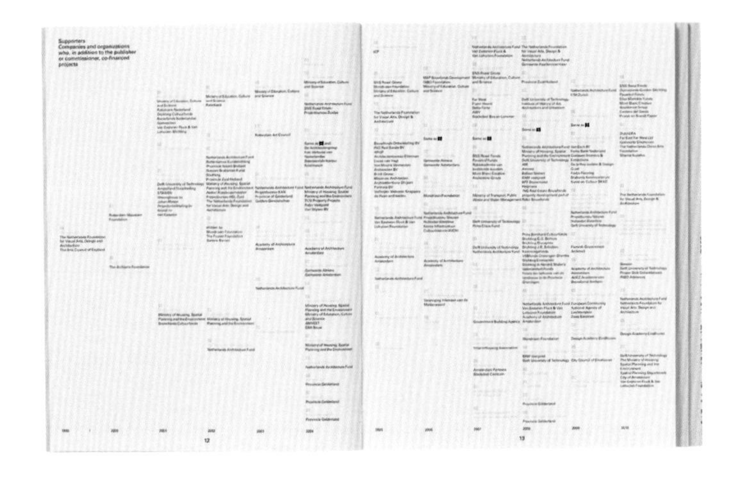

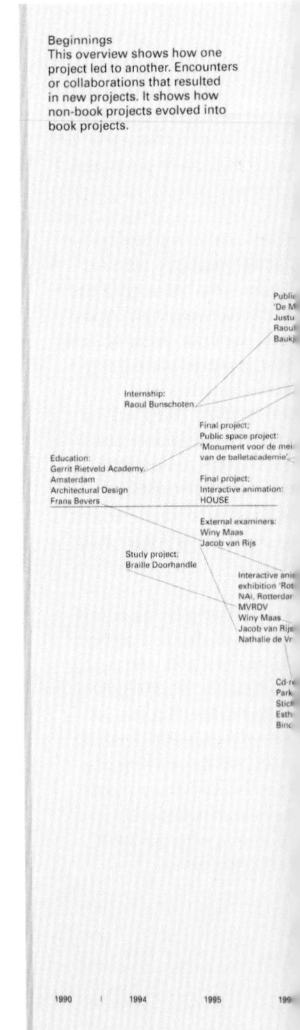

▶ 01

设计者：Joost Grootens
作品名：*I sweat I use no art at all*
细线用于说明项目与项目之间的关系

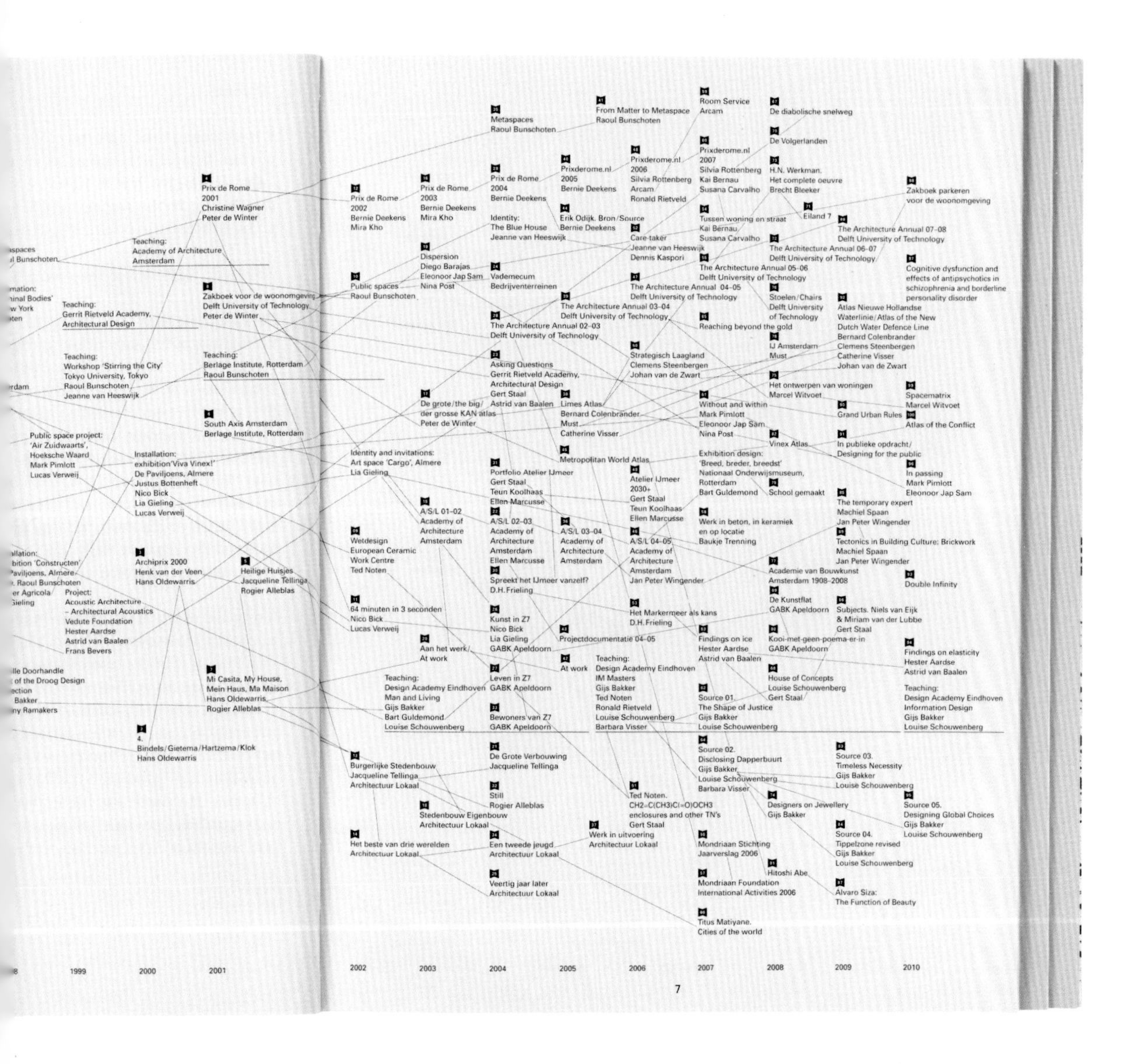

01.1.10
线性

这本432页关于戴姆勒股份公司的历史的书包含了三个部分：卡尔弗里德里希·奔驰、戈特利布·戴姆勒和威廉·迈巴赫的详细传记，戴姆勒公司从1883年到2011年的年表，20个关于汽车的故事、公司历史、外在品牌和产品及其内容方面的每一相关主题。一条银色丝带贯穿整本书，充当了包含详细说明和相关注释的时间轴与信息栏。

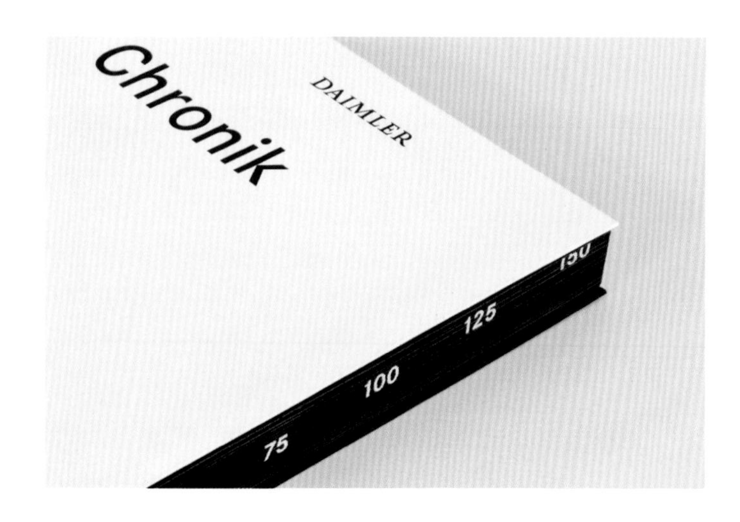

▸ 01
设计者：L2M3 Kommunikationsdesign
Daimler Chronik
125年的汽车史

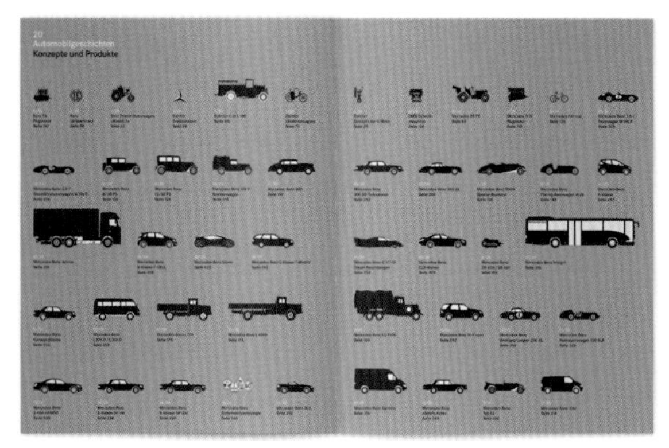
20
Automobilgeschichten
Konzepte und Produkte

20
Automobilgeschichten
Orte

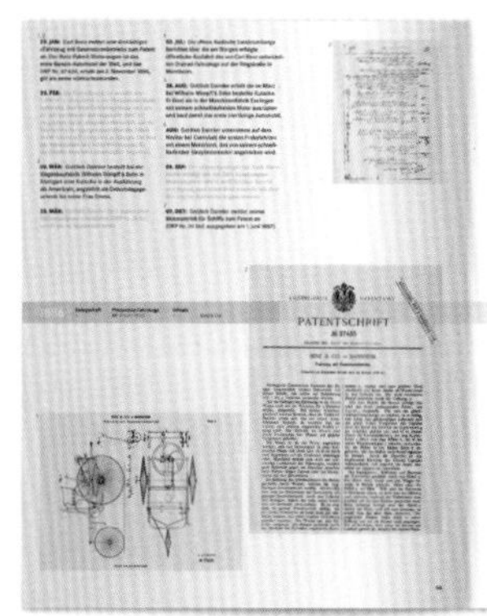

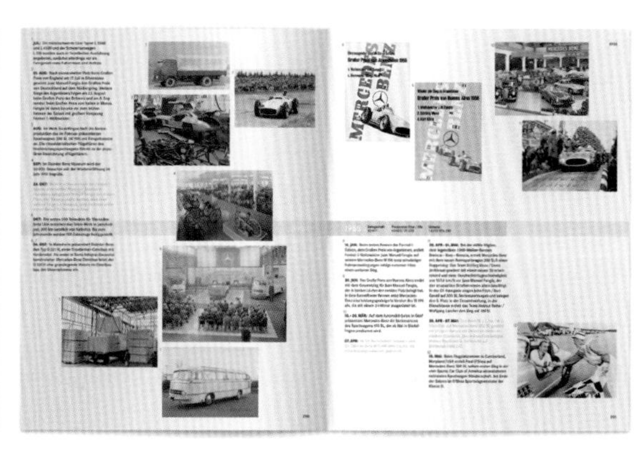

1897
Belegschaft
250
184
Produktion Fahrzeuge
256
26
Umsatz
0,666 Mio. M
Benz & Cie.
Daimler-Motoren-Gesellschaft
8, 9, 10
JUL: Benz & Cie. liefert das erste Fahrzeug des
Modells Break aus. Der offene Wagen für 12
Personen wird von einem 15 PS / 11 kW starken
Zweizylinder angetrieben. Wegen der ent-
gegengesetzt angeordneten Zylinder bezeichnet
Carl Benz diesen ersten Boxer-Motor als
Contra-Motor. Eine kleinere Ausführung mit
13
03. NOV: Das von dem ungarischen Erfinder
David Schwarz konstruierte Ganzmetallluftsch
das mit einem 12 PS / 9 kW starken Daimler-
Motor ausgerüstet ist, startet auf dem Tempel
hofer Feld zu seiner Jungfernfahrt. Bei der
Landung wird dieses erste Starrluftschiff der
Welt zerstört, der Pilot Ernst Jagels kommt
mit leichten Verletzungen davon.

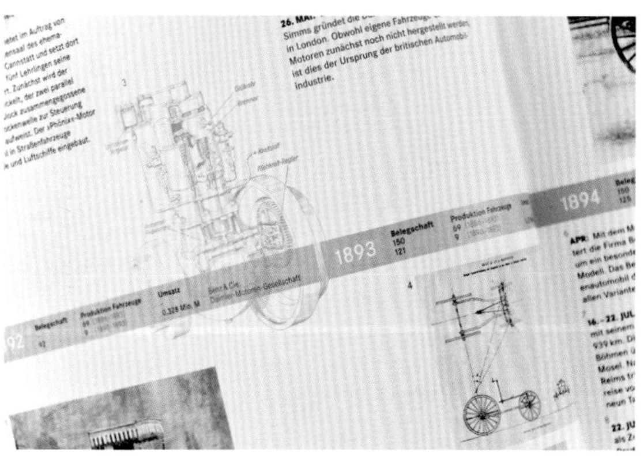
in London. Obwohl eigene Fahrzeuge
Motoren zunächst noch nicht hergestellt werden,
ist dies der Ursprung der britischen Automobil-
industrie.
1893
Belegschaft
150
121
Produktion Fahrzeuge
69
9
Umsatz
0,328 Mio. M
Benz & Cie.
Daimler-Motoren-Gesellschaft
1894

01.1.10
线性

《冰封的梦想》——一本关于俄罗斯当代艺术的书籍，采用了节省空间的排版方法，其用两页时间表描述了自1963年至今的俄罗斯艺术界。因为信息量年复一年的变化，所以为了节省版面，只有年度性的大事件被罗列出来。该方法虽然紧凑，但却失去了时间感和事件的连续感。

《释放》——一本有关土耳其当代艺术的书籍，与《冰封的梦想》属于同一系列，其时间轴的设置超出4个页面，这意味着可以使用更传统的横向板式。每十年占用一个页面，使人们可以清楚了解每年的信息量。

《古希腊遗址》一书的特点是扩展出超过12页的大量、多重时间轴，从公元前3000年一直延续到公元1900年。这些时间轴覆盖着如此漫长且包含众多不同层面的时期及该时期内各式各样的信息，这种时间轴使用了一个灵活的系统，该系统依赖于每一时期内的信息量。由此可以看出，公元前3000年至公元前2000年与公元前480年至公元前460年占据相同的空间。这些信息分为三个水平线索：历史事件、遗址以及艺术文化活动。

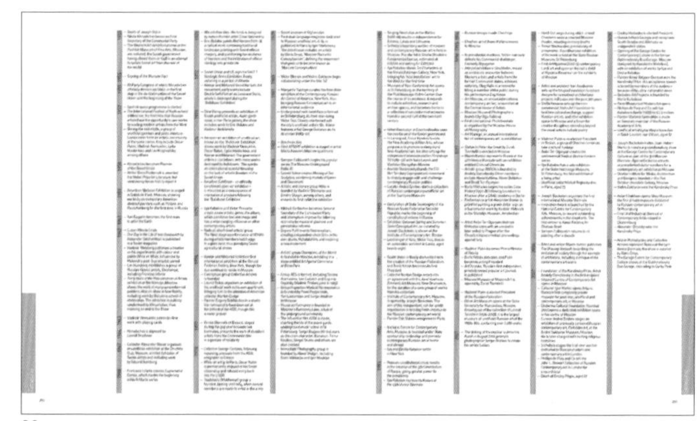

02

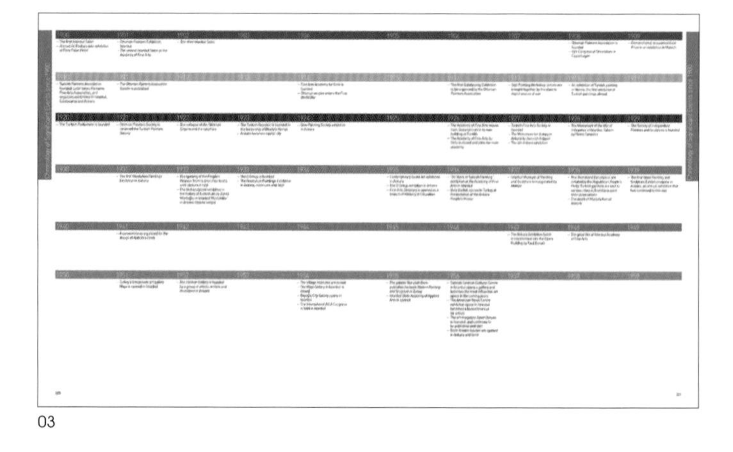

03

▸ 02

设计者：Struktur Design
作品名：*Frozen Dreams*
俄罗斯当代艺术
历史艺术时间轴

▸ 03

设计者：Struktur Design
作品名：*Unleashed*
土耳其当代艺术
历史艺术时间轴

▸ 04

设计者：Struktur Design
作品名：*The Sites of Ancient Greece By Georg Gerster*
古希腊历史时间轴

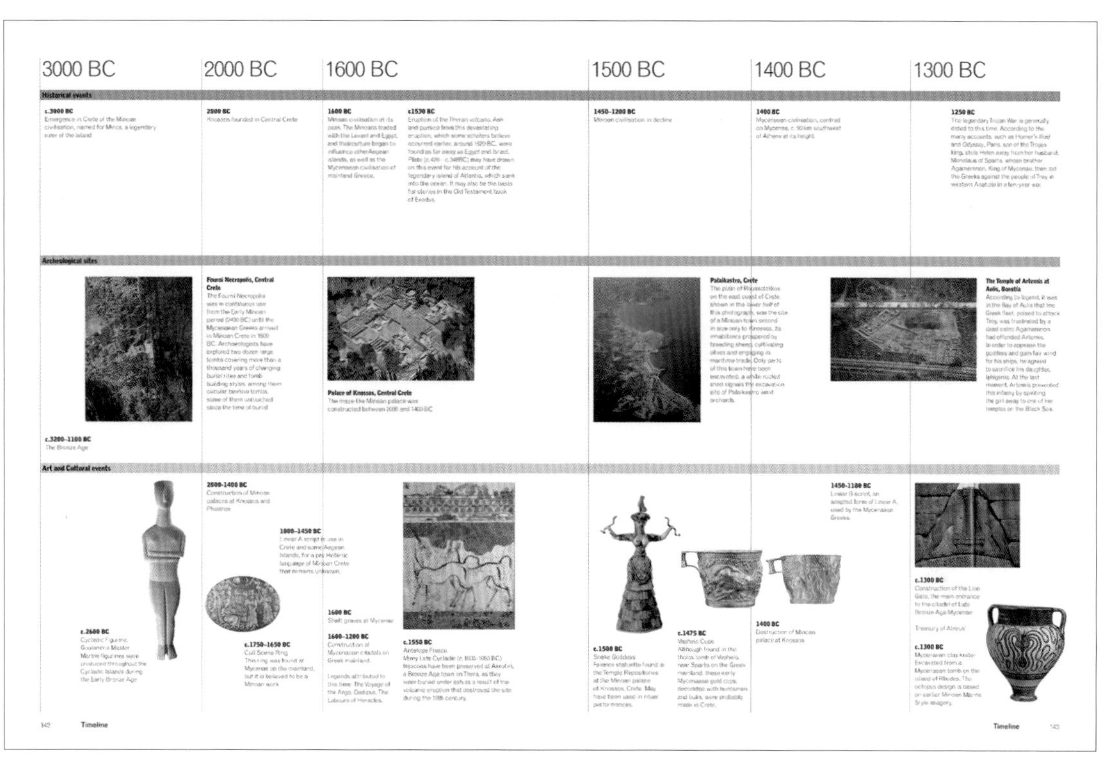
3000 BC
2000 BC
1600 BC
1500 BC
1400 BC
1300 BC
Historical events
Archeological sites
Art and Cultural events
Timeline
Timeline

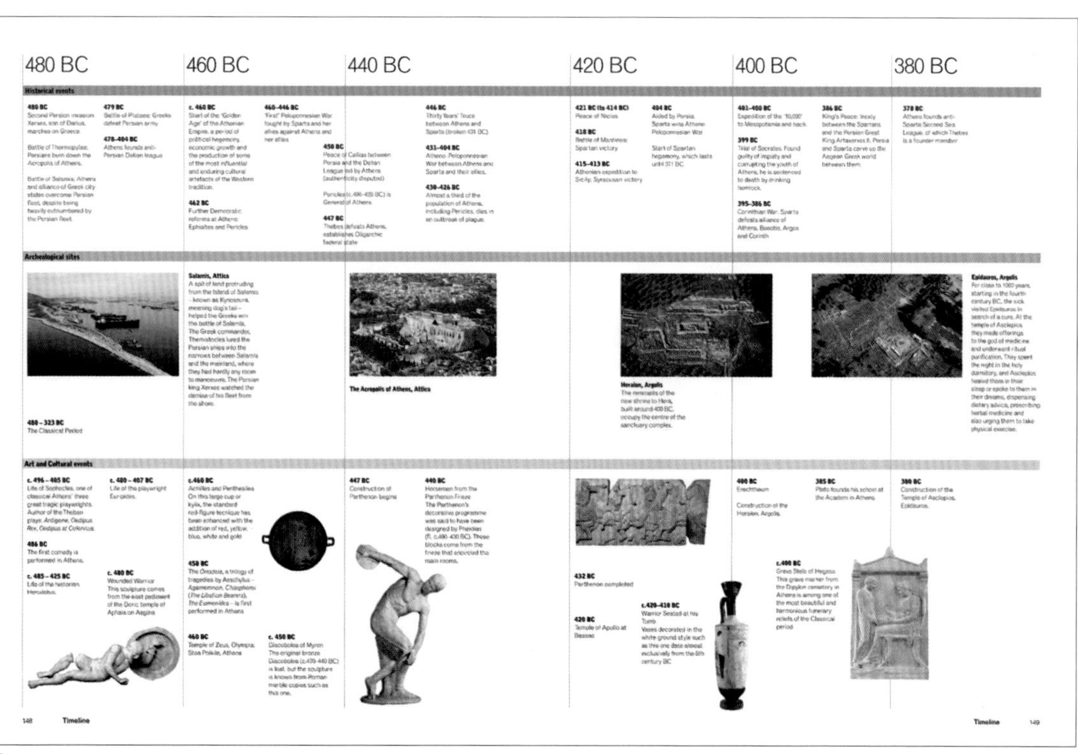
480 BC
460 BC
440 BC
420 BC
400 BC
380 BC
Historical events
Archeological sites
Art and Cultural events
Timeline
Timeline

04

01.1.11
数据尺度

《社交网络》是为2009年1月*Beef*杂志上的一篇文章制作的一项双面版式可视化设计作品，是对网络和社群的关系的设计。它显示50个主要的活跃社交网站，同时也显示了它们的用户数量（字体大小）及其启用时间（时间轴）。

▸ 01

设计者：Von B und C
Barbara Hahn，Christine Zimmermann
社交网络
*Beef*杂志

MySpac
235'000'000

Facebo
120'000'000

bo Windows Live

endster

70'000'000
5 Tagged
67'000'000
Orkut
63'000'000

ion Flixster

36'000'000
40'000'000
Netlog Bebo

kedIn
Odnoklassniki
29'000'000
24'000'000
imeem
V Kontakte
fm
Mixi
15'000'000
17'000'000
15'000'000
WAYN
Xiaonei
Buzznet
Nasza-klasa
Sonico
Geni
Hyves
MyYearbook
StudiVZ
Viadeo

2004 2005 2006 2007 2008 2009

Social Networking Websites

50 Internetcommunities
mit Launchjahr und Userzahl
Quelle: http://en.wikipedia.org/wiki/
List_of_social_networking_websites
(Stand: 06.03.09)
Visualisierung: Hahn und Zimmermann

01.1.12
尺度感

《一百万》是一份纲要的随机统计，其设计灵感来自作者亨德里克·赫茨伯格（Hendrik Hertzberg），他想使一组具有新闻价值的大型抽象数字变得更具体。纽约“Think”工作室设计出了修订后的最新新版本，最后的效果就是这一百万个点。与随机数字相对应的注释提供了有助于强化这一概念的不寻常的社会信息。

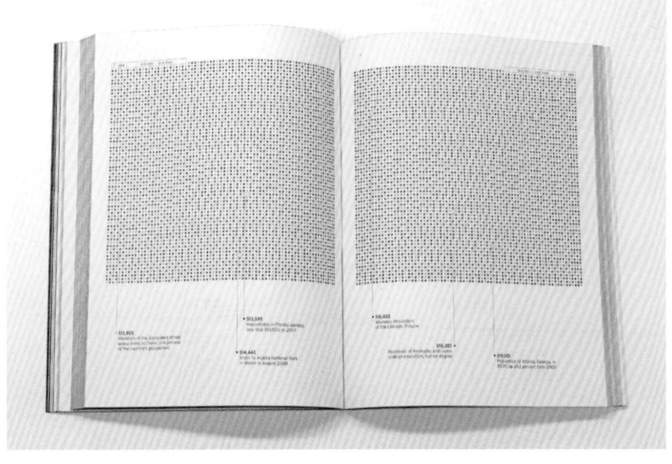

01

设计者：Think Studio, NYC
Herb Thornby, John Clifford

作品名：*One Million*
By Hendrik Hertzberg

艾布拉姆斯影像出版社

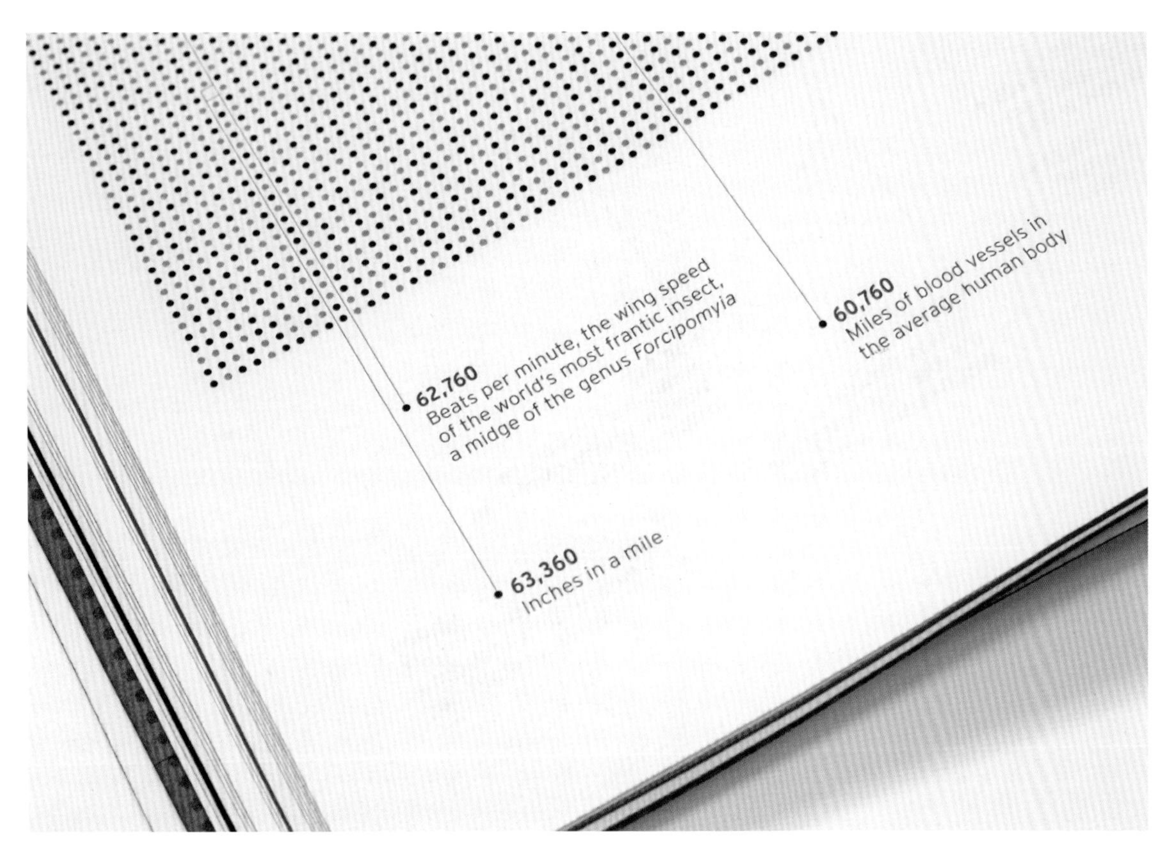
62,760
Beats per minute, the wing speed of the world's most frantic insect, a midge of the genus Forcipomyia
60,760
Miles of blood vessels in the average human body
63,360
Inches in a mile

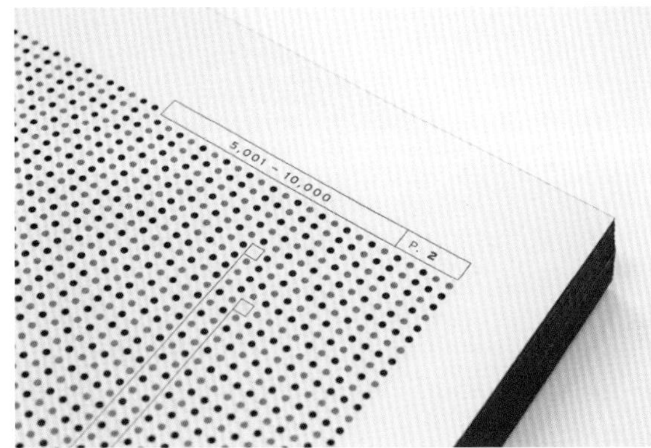
5,001 – 10,000
P. 2

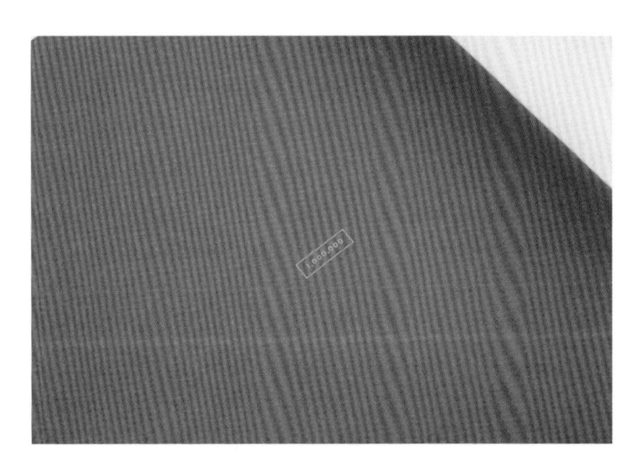

01.1.13 时效性

《有自信就有好结果》是一项旷日持久且有时效性的装置艺术作品，由一万根香蕉担当主角，生动地说明时间的流逝。作品以黄色的香蕉作为背景，未成熟的绿色香蕉用来组成文字。在数天以后，当绿色的香蕉成熟并且变黄，字母便会消失。随着时间的推移，黄色的背景会变成褐色而字母将仍保持黄色，直到数天以后，一切都会变得一塌糊涂。

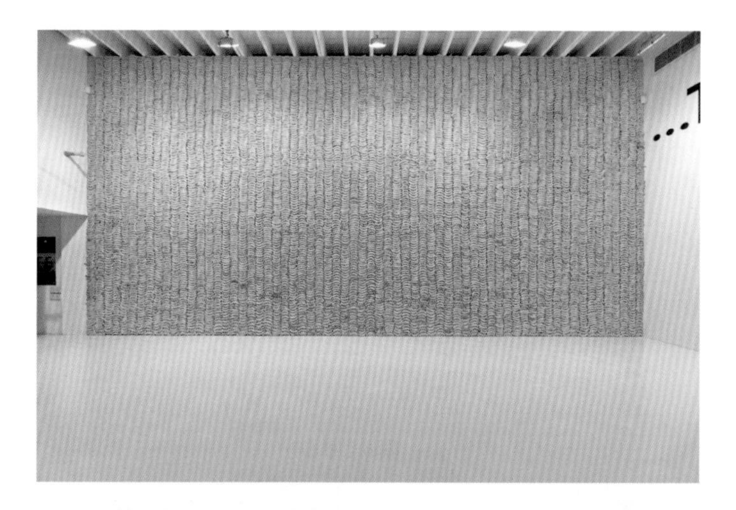

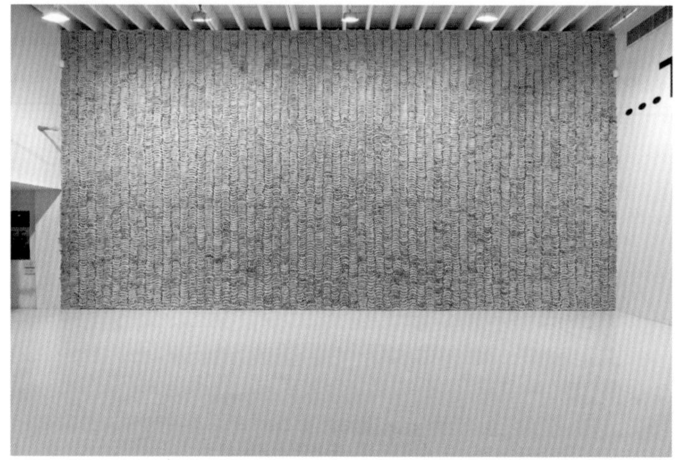

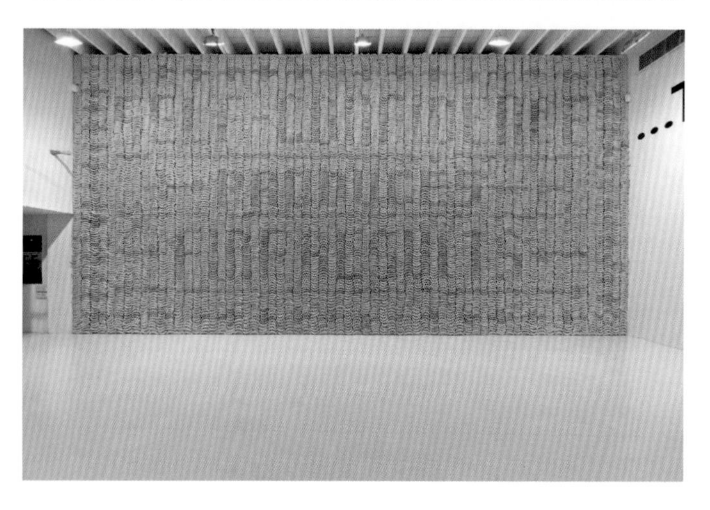

01

设计者：Sagmeister Inc.
Richard The, Joe Shouldice
艺术指导：Stefan Sagmeister
作品名：*Self-Confidence Produces Fine Results*
纽约戴伊奇项目一项基于时间的"陈述"

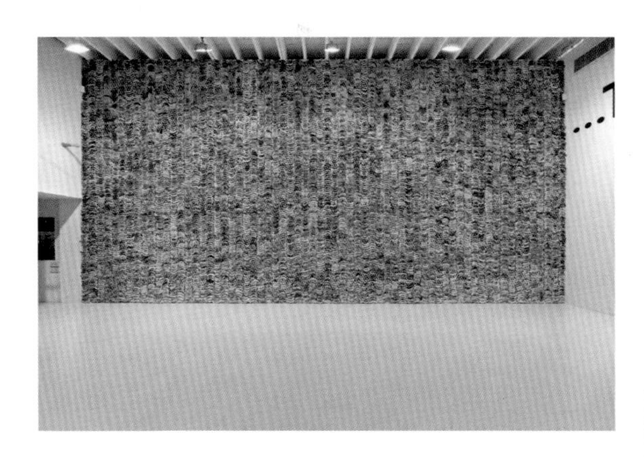

01.2
计算方法

01.2.1
时间间隔

这张由卡琳·冯·奥普特塔（Karin von Ompteda）为为期一天的“可视化研究研讨会”设计的海报，其时间轴通过图表进行表达，用以表现时间的流逝。

海报中时间栏沿左侧向下延伸，显示了这一天中展开讨论的每一分钟。半圆弧被用来形象化地说明每一次讨论的持续时间，根据每个人讨论的时间长短来增加圆弧的大小，从而使日程表更易阅读。左侧边缘的时刻表以较大号印刷字体代表一系列讨论的开始，较小字体则代表在会议中进行的不同的主题阶段。小半圆代表每次讨论的时间长短，而大半圆则将会谈中的相关发言者和所有讨论联系在一起。

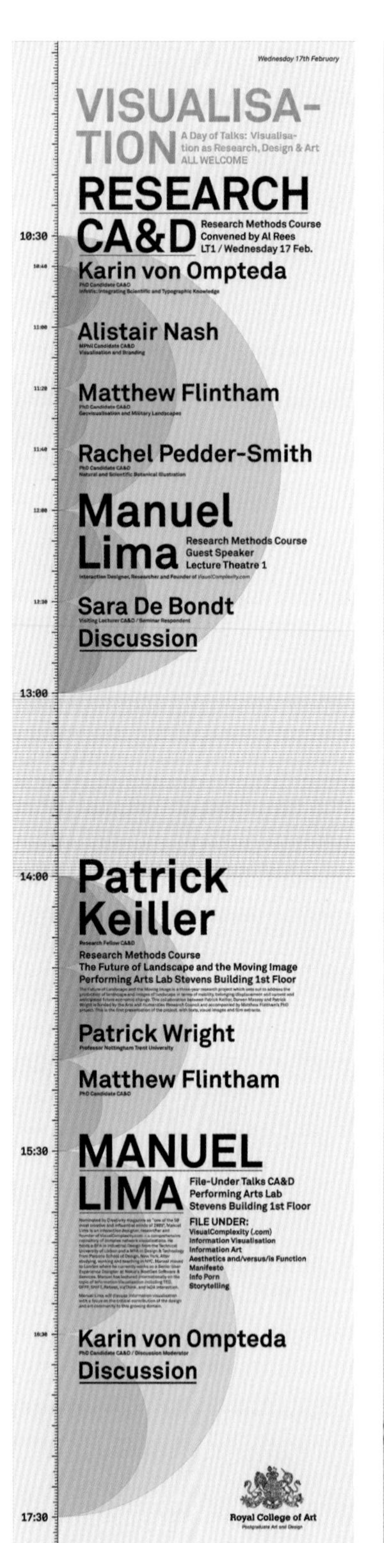

▶ 01

设计者：Karin von Ompteda
作品名：*Visualisation Research*
“可视化研究研讨会”海报

VISUALISA-
TION
A Day of Talks: Visualisa-
tion as Research, Design & Art
ALL WELCOME
RESEARCH
CA&D
Research Methods Course
Convened by Al Rees
LT1 / Wednesday 17 Feb.
10:30
10:40
11:00
11:20
Karin von Ompteda
PhD Candidate CA&D
InfoVis: Integrating Scientific and Typographic Knowledge
Alistair Nash
MPhil Candidate CA&D
Visualisation and Branding
Matthew Flintham
PhD Candidate CA&D
Geovisualisation and Military Landscapes
Rachel Pedder-Smith
PhD Candidate CA&D
Natural and Scientific Botanical Illustration
Manuel
Research Methods Course
Guest Speaker
Lecture Theatre 1

01.2.2
时间流逝

《Blanka-Helvetica字体50周年》（*Blanka-Helvetica 50*）是为庆祝Helvetica这一遍及全球的字体诞生50周年而举行的展览所设计的海报、图形及邀请函。50位设计师被邀请参加本次展会，每位设计师为1957——2007年这50年中的其中1年配了插图。布德（Build）设计的1969年描绘了尼尔·阿姆斯特朗（Neil Armstrong）说出的有历史意义的话“一小步……”。他们设计的展览邀请函将50年这一时代换算为秒、分、时和日。

一份由“Stapelberg & Fritz”制作的社论特刊记载了大众帕萨特的历史，并将多年来每一时期的汽车照片与超大号数字数据相结合。

“Design Project”设计公司为佛捷歌尼公司的新系列“Arcoprint”环保纸张设计的宣传海报由两种色彩搭配方案制作而成，并由不同克重的纸张（90~350 g/m²）来印刷突出了其特色。

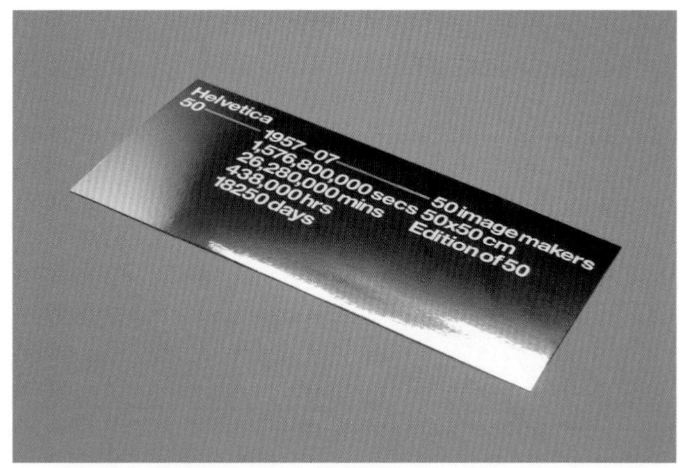

01

02

03

▶ 01
设计者：Build
作品名：*Blanka-Helvetica 50*
展览邀请

▶ 02
设计者：Build
作品名：*Blanka-Helvetica 50*
海报

▶ 03
设计者：Stapelberg & Fritz
作品名：*The New Passat Driving Experience*
EVW帕萨特杂志

▶ 04
设计者：Design Project
作品名：*Arcoprint by Fedrigoni*
促销纸样品

Arcoprint —
New limited
edition paper
sampler
Leeds
14 07 2010
Fedrigoni invite you to the launch of a new limited edition paper sampler, demonstrating the usability of Arcoprint — an extra white premium quality uncoated material.
Product launch:
The Cross Keys, Leeds.
Wednesday 14 July,
6 — 9pm.
RSVP by emailing info@fedrigoni.co.uk and receive a free food/refreshment voucher.
90

04

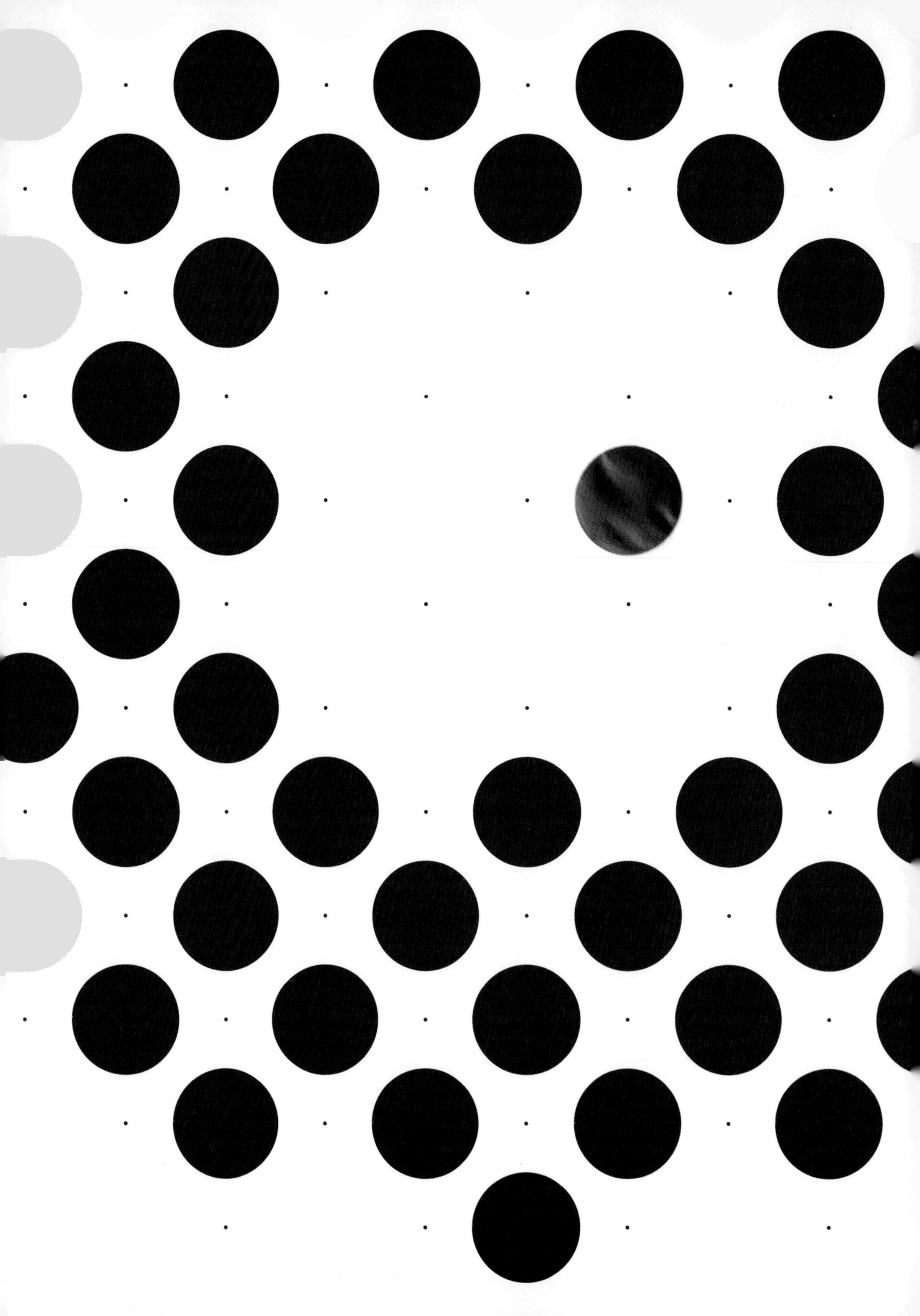

02. 数据

02.1 量化

02.1.1 比例划分 044
02.1.2 交叉与网络 046
02.1.3 铁一般的事实 048
02.1.4 自然形式与顺序 050
02.1.5 规模与活动 054
02.1.6 可视化分析 056
02.1.7 基于时间 062
02.1.8 社交混杂 064

02.2 统计

02.2.1 触摸屏 066
02.2.2 数据大爆炸 068
02.2.3 尺度的概念 070
02.2.4 革命 072
02.2.5 视觉分析 074
02.2.6 数据丰富 076
02.2.7 方向性 078
02.2.8 工作流程 080

02.3 报告与说明

02.3.1 饼图 082
02.3.2 柱状图 084
02.3.3 无图表 086

02.1
量化

02.1.1
比例划分

在《奥迪邂逅杂志》中，实色色块代表奥迪A4车型结构所使用的各种构成要素。每种材料的质量（单位：千克）及其在总构架中的百分比都一一列出。不同材料的有限数量表明了该车辆的可回收利用性。

《奥迪A8手册》的特色在于其章节分隔符印制于较厚的白色绘图纸上，其中倒置的章节编号采用纯白色墨水标注。与《邂逅》杂志一样，该手册以汽车材质结构的色调百分比为特色。

01

▸ 01

设计者：Stapelberg & Fritz
作品名：*Encounter*
奥迪周边杂志

▸ 02

设计者：Stapelberg & Fritz
作品名：*Audi A8*
奥迪A8宣传册

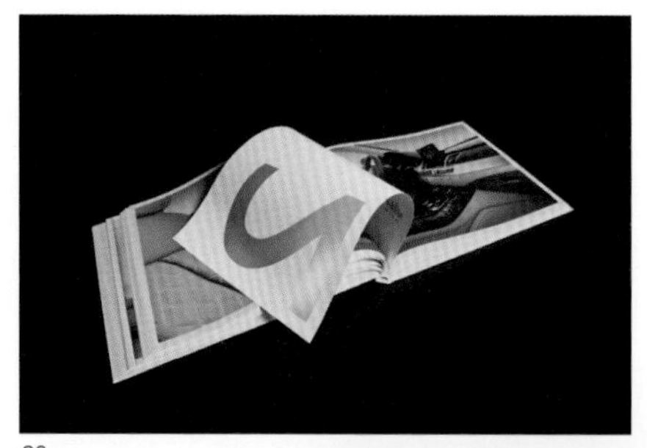

02

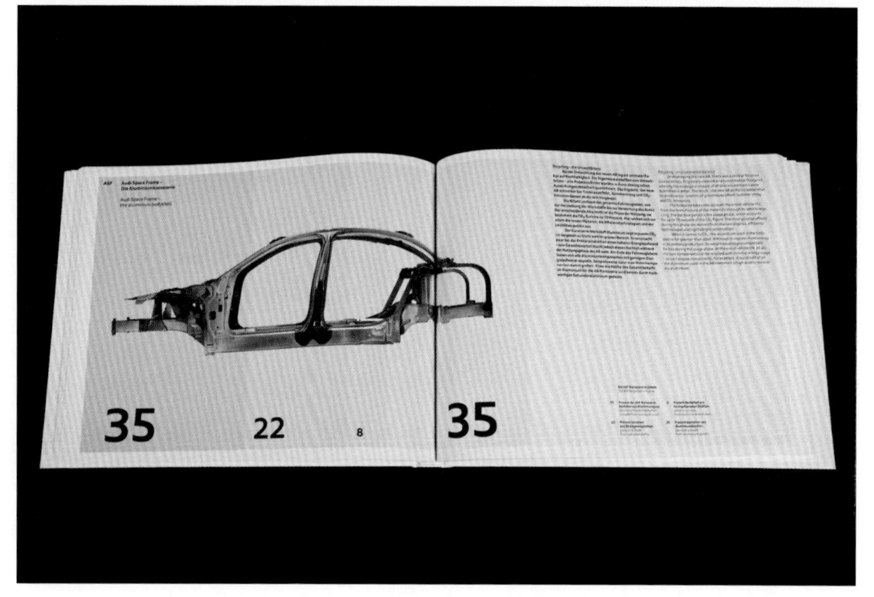

02.1.2
交叉与网络

数学理论中的多面体被认为是几何上最复杂、美学上最迷人的结构形式，在《理论与实践中的真正魔法》4-21中列举的多面体其实是一切普遍理论中心的一项代数形式。因为其根轴系向量位于八维欧氏空间，它们通常被称为“E8”。在维度结构内，它所包含的维度创建出复杂性惊人的248个匀称格状结构，当它在时空中扭转、折叠时，该结构可预测宇宙中所有已知的粒子及作用力。这张限量版的海报为黑色丝网印刷，在干净的聚丙烯板上采用23克拉金箔及粉末通过烫金工艺制作而成。

《纽约》杂志关于00年代问题的插图和信息图表是针对过去十年社交网络的发展所作的评论。每一线条代表的在Facebook（黑色）、Myspace（黄色）、LinkedIn（紫色）、Twitter（红色）、Bebo（绿色）及Ning（蓝色）上注册的120 000个用户。

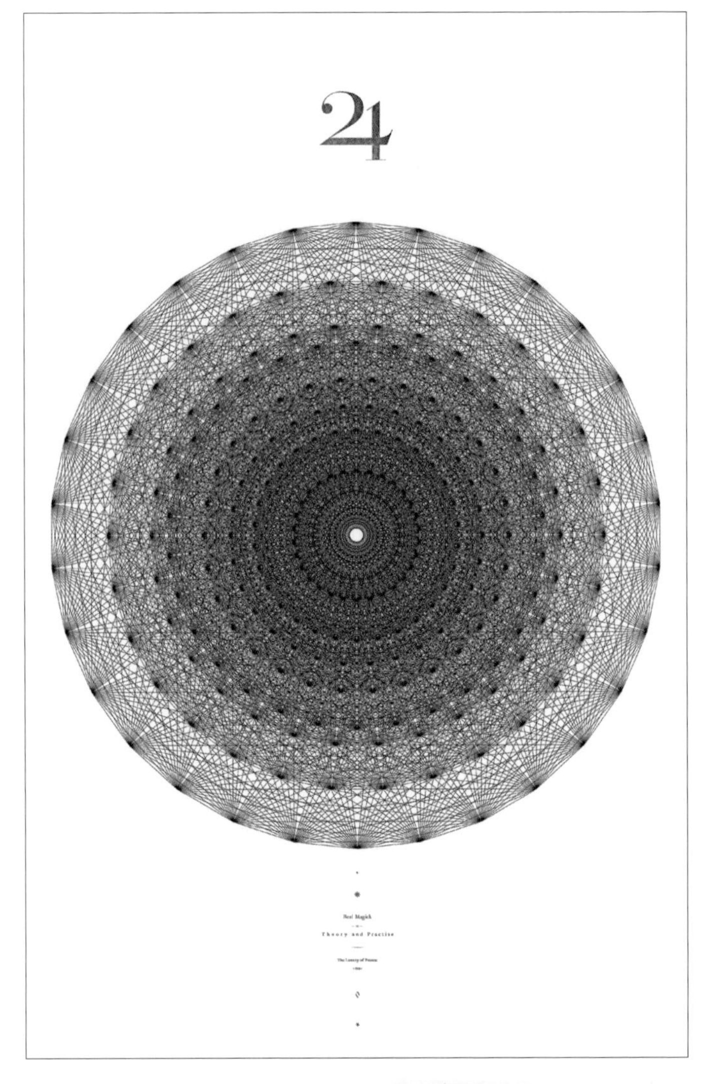

01

▸ 01

设计者：The Luxury of Protest
作品名：*Real Magick in Theory and Practise*
限量版海报

▸ 02

设计者：Studio8
作品名：*The 00s Issue*
《纽约》杂志

02

02.1.3
铁一般的事实

《两点》是一张为广岛和长崎遭受核袭击61周年制作的纪念海报。这张海报以柔和的图形来表现忧郁性质的主题：单一的色彩、极小的图形、统计信息列表以及类型要素的空间布局。这两个点用手工绘制于白色亚光漆标志物上，用来与玻璃样的超光滑塑料介质形成对比。这些点代表许多不同的概念：这两起轰炸事件，我们/他们，核裂变……它们象征着人类在轰炸中的悲剧，并附带显著的粗糙处理与平滑处理的对比。

《A-B和平与恐惧》和《爱与恨的计算美学》双面海报，显示的是对联合国192个成员国进行的地缘政治调查。A面显示的是和平的措施，B面显示的是对恐怖行动的对策。针对A、B面的每项措施，把表分为三个环（针对和平的三项单独索引以及针对恐怖的三项单独索引），这些环是从地缘政治领域的研究人员个人量化方式测定。和平与恐怖行动测量的数量变化表现为线条粗细变化：细线=低值，粗线=高值。

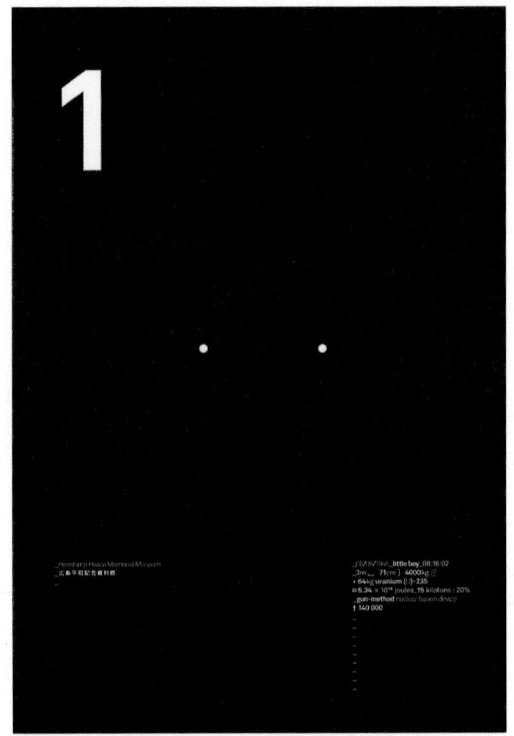

01

02

▸ 01
▸ 02

设计者：The Luxury of Protest
作品名：*2Dots*
为广岛/长崎战争博物馆设计的限量版海报

▸ 03
▸ 04
▸ 05

设计者：The Luxury of Protest
作品名：*A-B-peace & Terror ect.*
《爱与恨的计算美学》限量版海报

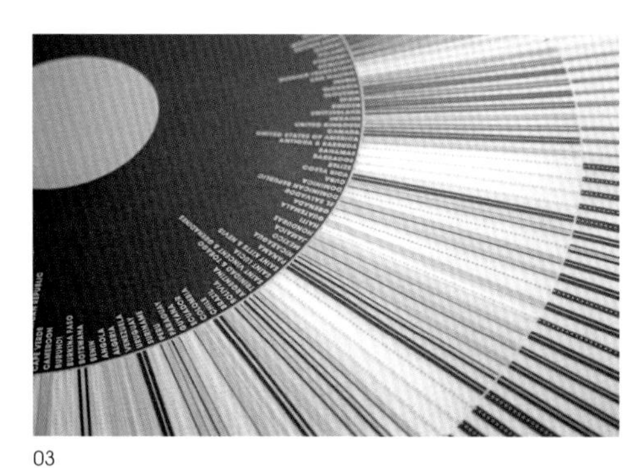

03

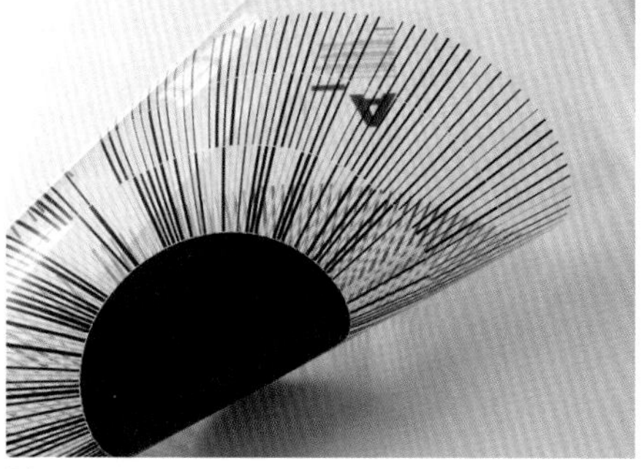

04

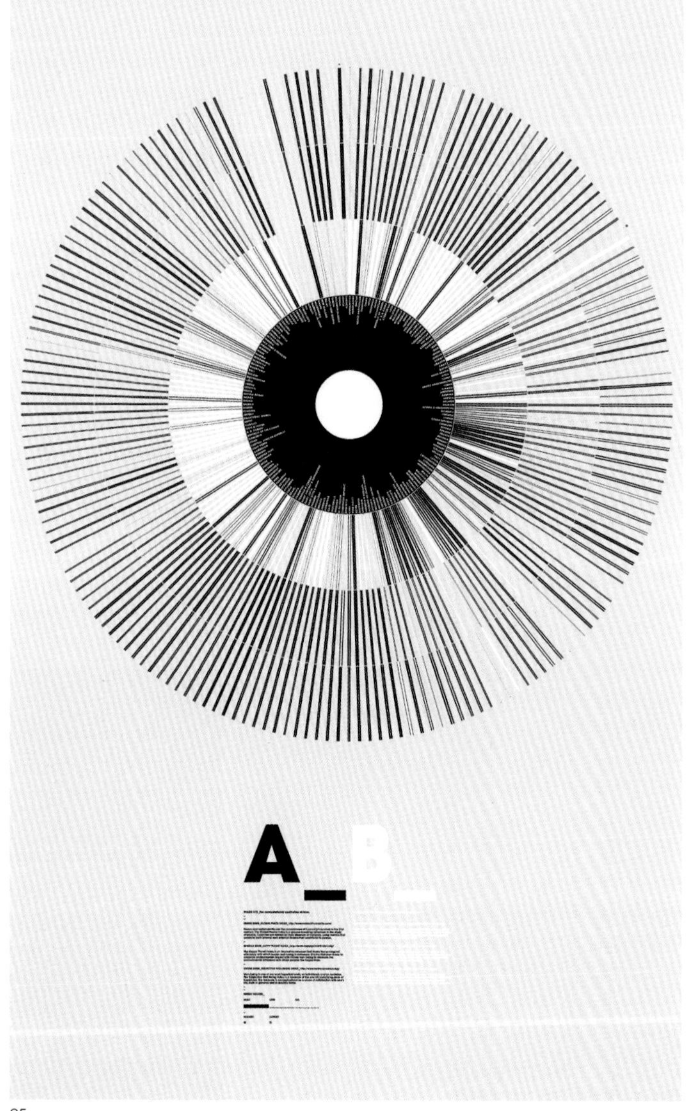

05

02.1.4
自然形式与顺序

《数学幻想宇宙》中的图形利用生成的“Python”代码及按照对数螺旋线排列的0到100 001的地图编号创建而成。螺旋形式由从中心（0）向外缘（100 001）进行数字分布的圆的黄金角子张力决定。这一模式经常被发现存在于自然界中（比如植物），并且自阿基米德时代以来就有所记载。螺旋形式显示的是基本数字的视觉关系及数学方程式的美感。

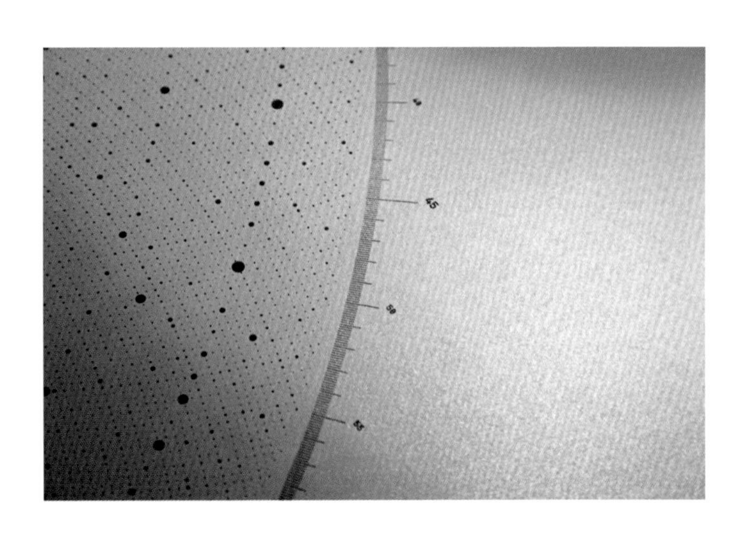

01

设计者：The Luxury of Protest

作品名：*Maths Dreamed Universe*

自然界中元素对其自身进行排列的方式的定量可视化

MATHS
DREAMED UNI – 0 to 100,001
VERSE.

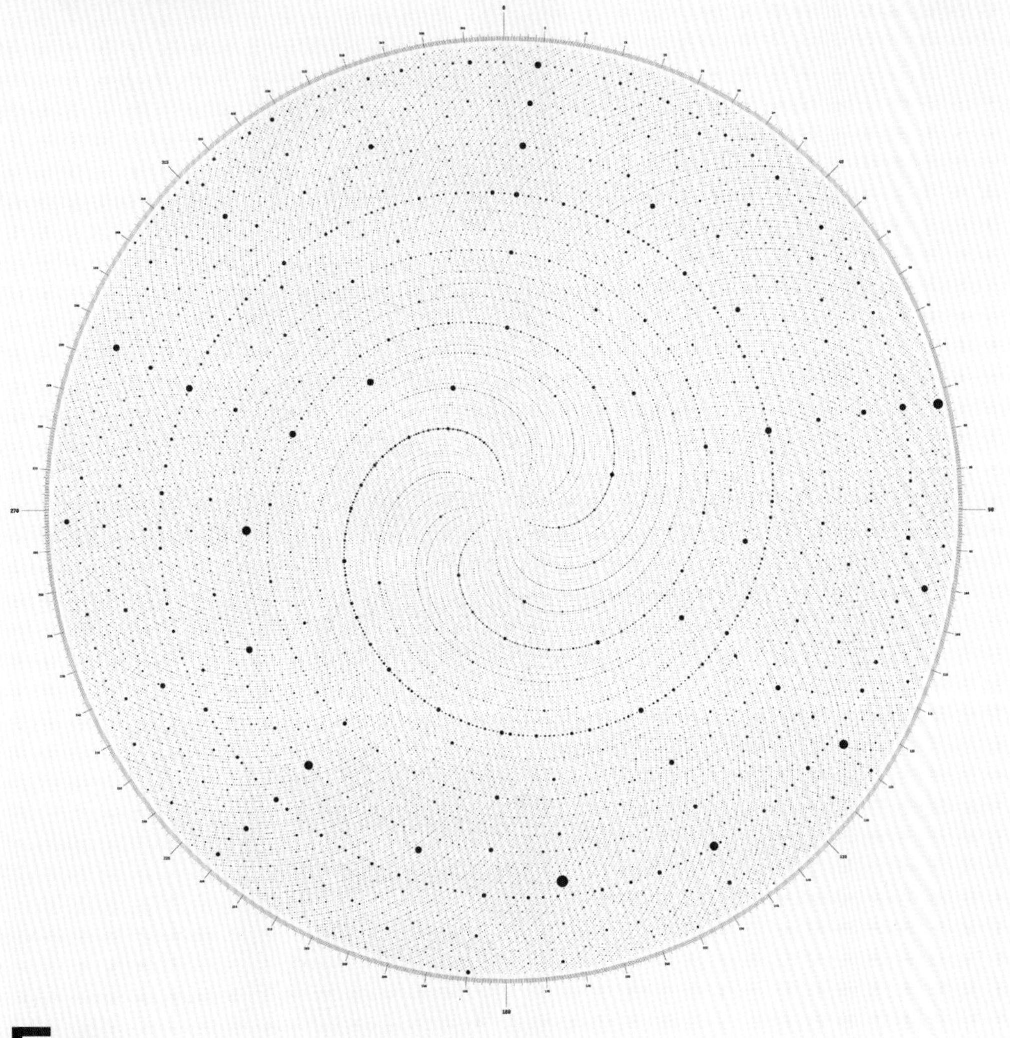

Eq:
$r=\sqrt{n},\ \theta=(2n\pi)/\phi^2$

02.1.4
自然形式与顺序

《照此下去》是为总部位于圣弗郎西斯科的慈善组织——雨林行动网（RAN）设计的一本小册子，他们与摄影师贾尔斯·雷维尔合作，目的在于增强人们对亚马逊雨林遭人类破坏的认识。

每本小册子均由森林管理委员会认证的一种纸张制作而成，从封面开始折叠成12页的风琴状小册子，达到最大化使用和最小化浪费。

小册子的每一次展开都呈现关于热带雨林的重要数据，且数字超出页面，与剩余的文本和精美的图片形成极大的反差。

海报《2060》的设计再次与吉尔斯·雷维尔合作，采用一幅精妙叶脉摄影图像搭配着醒目大尺度数字标题。

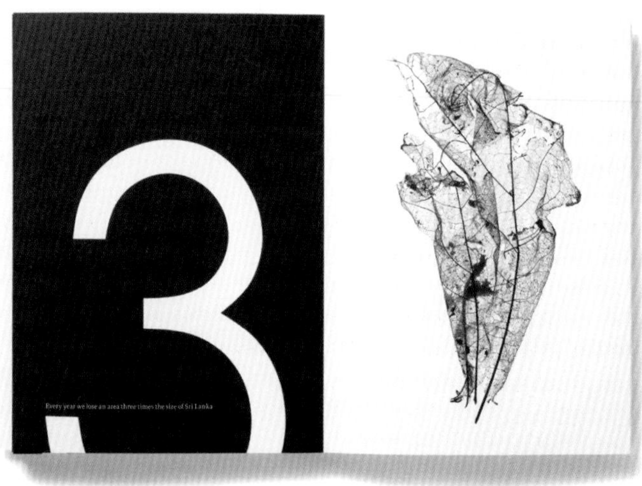

02

▸ 02

设计者：Studio8
作品名：*At This Rate*
雨林行动网（Rainforest Action Network）宣传册

▸ 03

设计者：Studio8
作品名：*2060*
雨林行动网海报

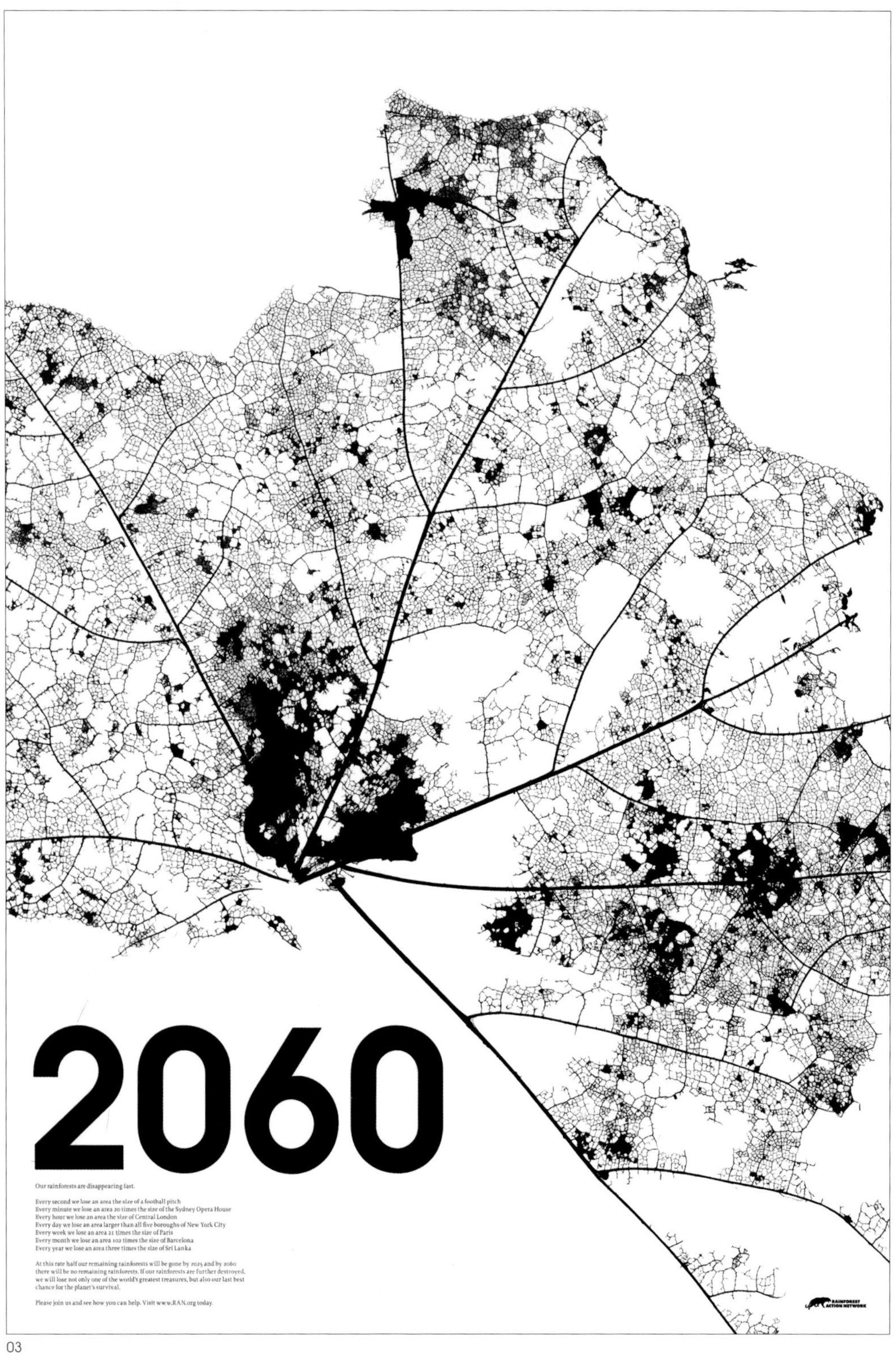
2060
Our rainforests are disappearing fast.
Every second we lose an area the size of a football pitch
Every minute we lose an area 20 times the size of the Sydney Opera House
Every hour we lose an area the size of Central London
Every day we lose an area larger than all five boroughs of New York City
Every week we lose an area 21 times the size of Paris
Every month we lose an area 102 times the size of Barcelona
Every year we lose an area three times the size of Sri Lanka
At this rate half our remaining rainforests will be gone by 2025 and by 2060 there will be no remaining rainforests. If our rainforests are further destroyed, we will lose not only one of the world's greatest treasures, but also our last best chance for the planet's survival.
Please join us and see how you can help. Visit www.RAN.org today.
RAINFOREST ACTION NETWORK

03

02.1.5
规模与活动

《柏林2007年营业额》对柏林文化产业及设计经济的不同方面进行可视化设计，它对比了2007年柏林的整体经济（G）、整个集群（C）、文化经济（K）及设计经济（D）。

《柏林设计场景》以设计场景中的83个地点为特色。每一类别——工作室、机构、博物馆等均通过一个彩色的星星来进行项目识别。每一地点均由一个数字来表示，其社区附近则使用彩色线来进行标志，这样便生成了相关区域的代表性网格。网格越密，则该区域存在的场地越多。场地的位置形成了不同区域特定的颜色和密度模式。

《柏林集群的发展》阐述的是2000——2007年创意产业的增长。浅灰色的标题参照于2000年，黑色标题参照于2007年。垂直刻度显示的是公司的数量，而水平刻度代表的是公司的规模。部门名称的垂直移动和向右移动代表行业的增长及其规模的增加。

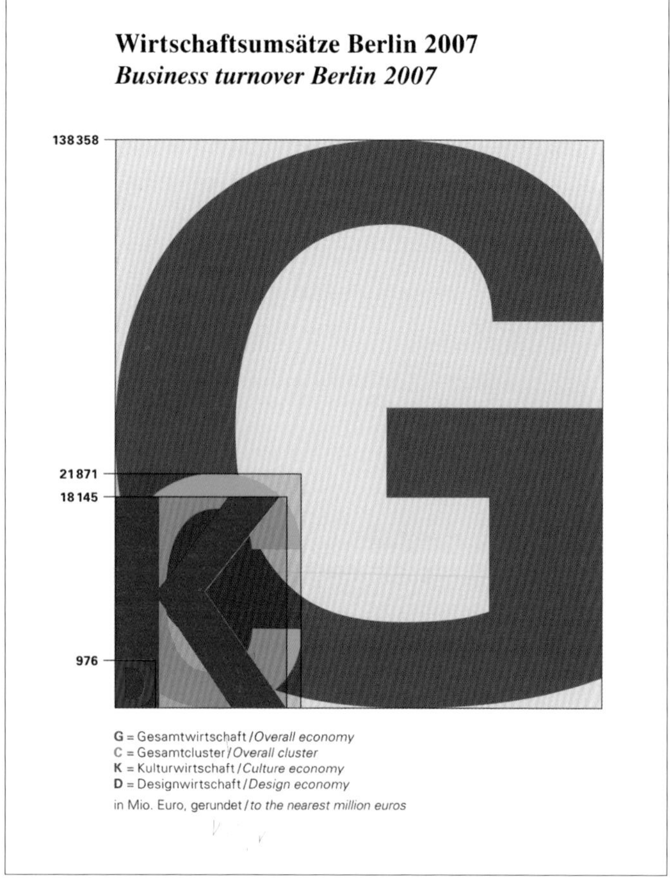

01

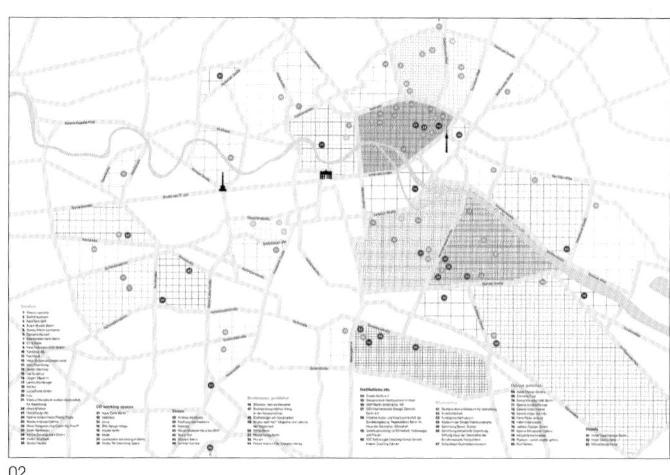

02

▸ 01

设计者：Von B und C
Barbara Hahn, Christine Zimmermann
作品名：*Business Turnover 2007*
《Form》杂志

▸ 02

设计者：Von B und C
Barbara Hahn, Christie Zimmermann
作品名：*Design Scene Berlin*
《Form》杂志

▸ 03

设计者：Von B und C
Barbara Hahn, Christie Zimmermann
作品名：*Development of the Berlin cluster Communication, media, creative industries form 2007-07*
《Form》杂志

ntwicklung des Clusters Kommunikation, Medien, Kreativwirtschaft Berlin von 2000–2007
evelopment of the Berlin cluster communication, media, creative industries from 2000–2007

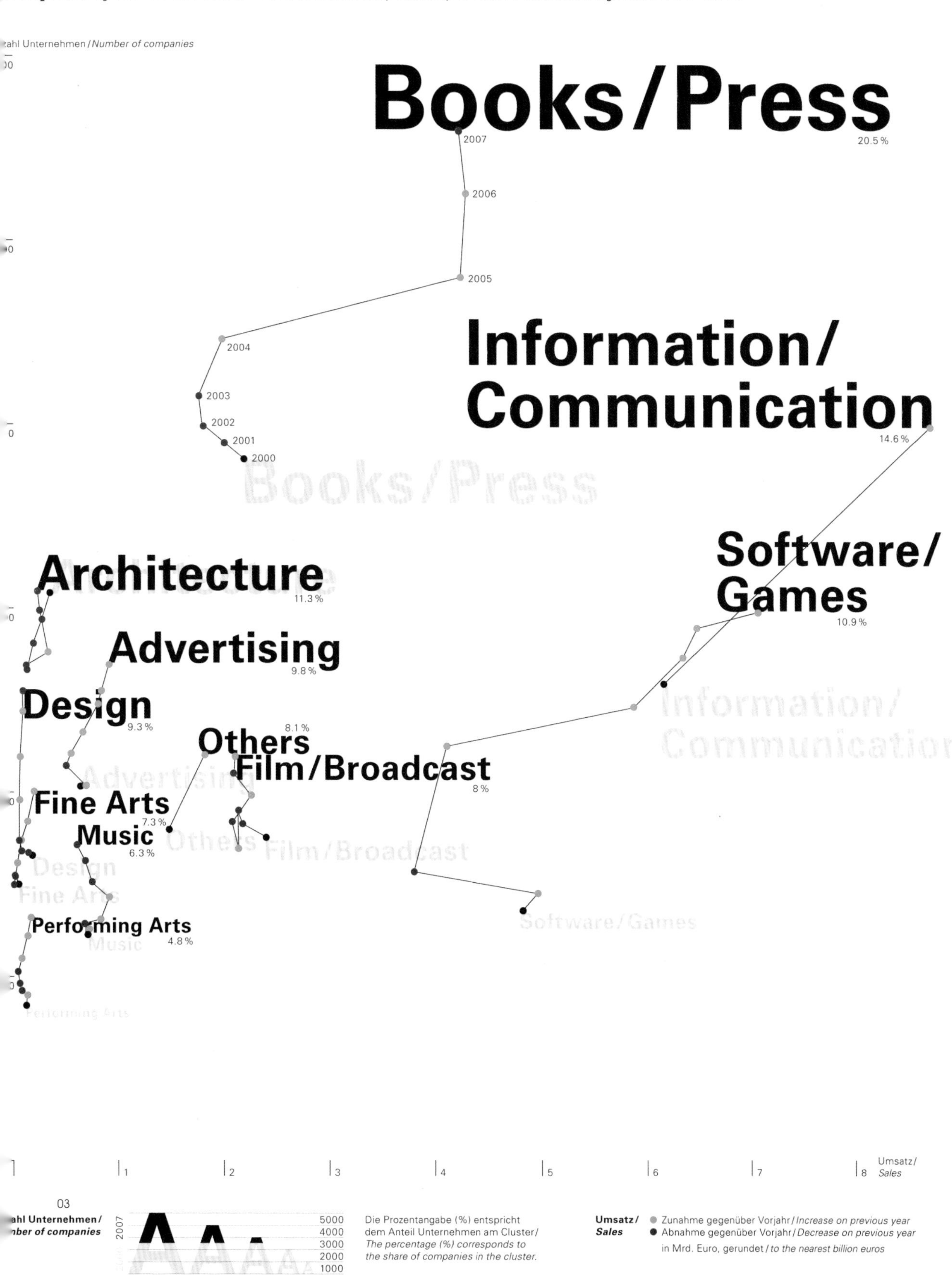

Die Prozentangabe (%) entspricht dem Anteil Unternehmen am Cluster / *The percentage (%) corresponds to the share of companies in the cluster.*

Umsatz / *Sales*
● Zunahme gegenüber Vorjahr / *Increase on previous year*
● Abnahme gegenüber Vorjahr / *Decrease on previous year*
in Mrd. Euro, gerundet / *to the nearest billion euros*

02.1.6 可视化分析

《医院中日常生活的视觉图谱》研究项目中，四项独立而创新的可视化设计得以开发，以帮助呈现伯尔尼大学附属医院患者看病过程中所选择的组织性和交际性子流程。最终在医院内可以更清楚地呈现工作流程和结构，并可以对其进行分析从而达到更有效的管理。

已对四项可视化子流程的数据进行了收集，用于医院质量管理部门的流程分析项目。90位出院病人对问卷调查作出了回应，该问卷调查旨在保证可视化的信息水平并回答关于患者康复的重要问题。

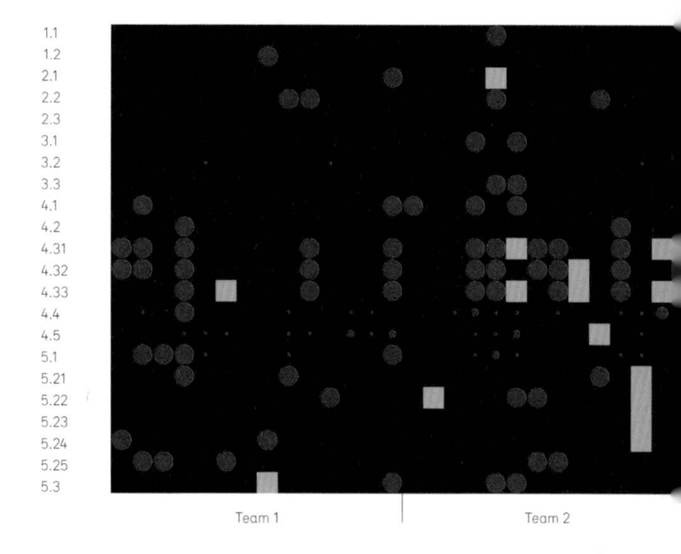

▸ 01

设计者：Von B und C
Barbara Hahn，Christine Zimmermann
作品名：*Visual Atlas of Everyday Life at the Hospital*
伯尔尼艺术学院/伯尔尼大学附属医院

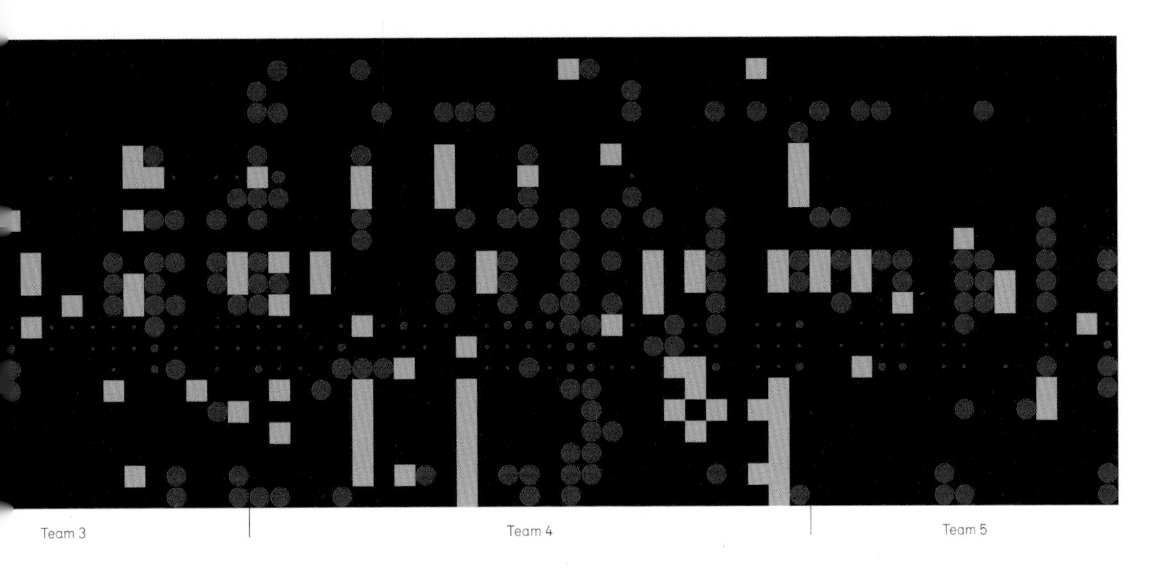

t informiert

mittelmässig informiert

schlecht informiert

sehr schlecht informiert

Daten nicht erhoben

02.1.6
可视化分析

在“女人的手机”这一项科学研究中《社交地图》显示的是7位女性（A~G）的社交网络和社交行为（面对面、电话、网络电话、电子邮件）。7位女性中每一位的所有通信伙伴均放射状地分布在她们周围。5个外圈上（每一圈表示一种通信形式）某一特定女性的所有通信形式均通过有颜色的笔画来突出。这些颜色说明的是与配偶（橙色）、家人（红色）、朋友（紫罗兰色）、熟人（紫色）、同事（蓝色）间的联系。笔画的粗细与相关人地理位置的近或远有关：同一城市（粗笔画）、同一国家（中等笔画）、同一世界（细笔画）。

《形容词的尺度》显示的是对66个对象进行的分析，这些物件由“女性手机”研究项目中的5位成员收集而来。就这些物体的49个相关特性（有机的、装饰性、小、柔轻软、光滑、有光泽、引人注目的、芳香的、精细、粗糙性，等等）进行了分析。阵列（形容词在X轴上，物件在Y轴上）通过三种色阶强度的色彩编码来说明某种特性是极易或部分，还是从未应用于某个物件。通过从紫罗蓝色到橙色这些浓淡不一的红色来对5位成员进行巧妙的区分。

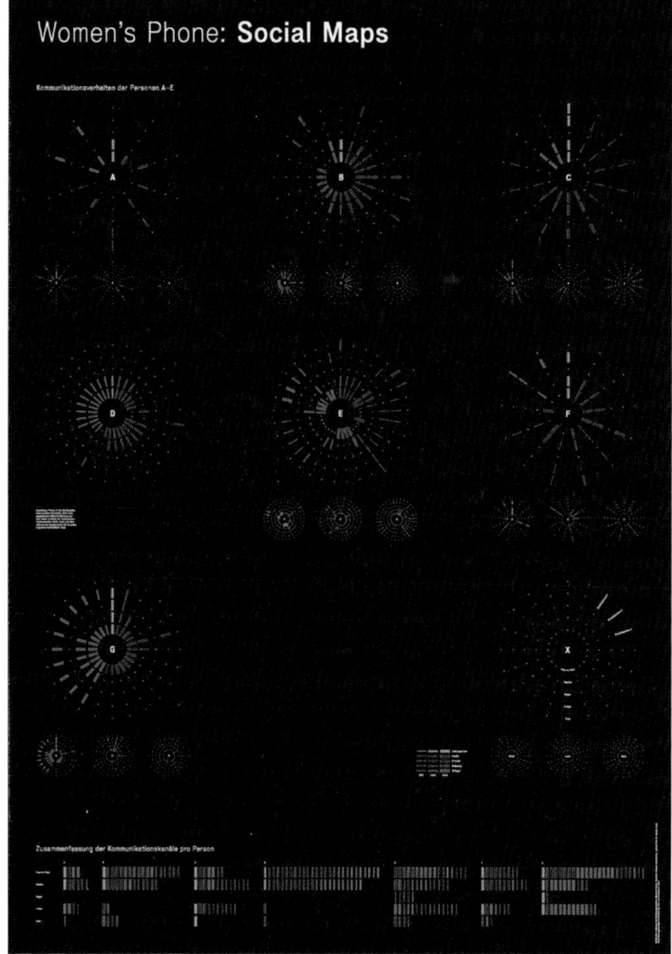

02

▸ 02

设计者：Von B und C
Barbara Hahn, Christine Zimmermann
作品名：*Women's Phone: Social Maps*
德国电信实验室

▸ 03

设计者：Von B und C
Barbara Hahn, Christine Zimmermann
作品名：*Women's Phone: Adjectival Scales*
德国电信实验室

▸ 04

设计者：Von B und C
Barbara Hahn, Christine Zimmermann
作品名：*Women's Phone: Adjectival Scales*
德国电信实验室

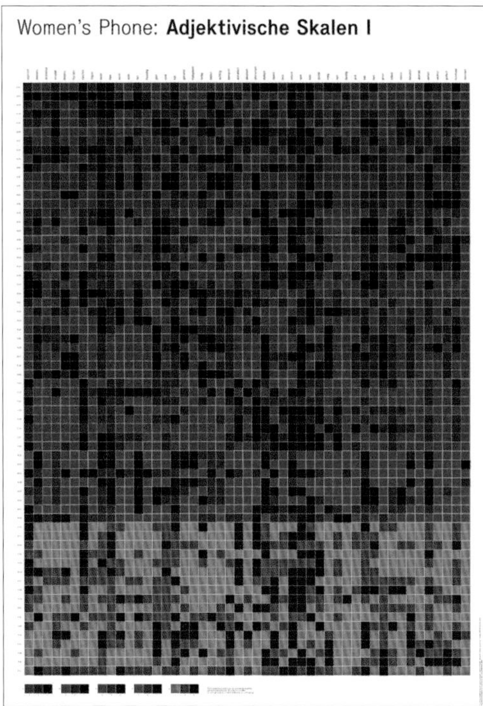
Women's Phone: Adjektivische Skalen I

03

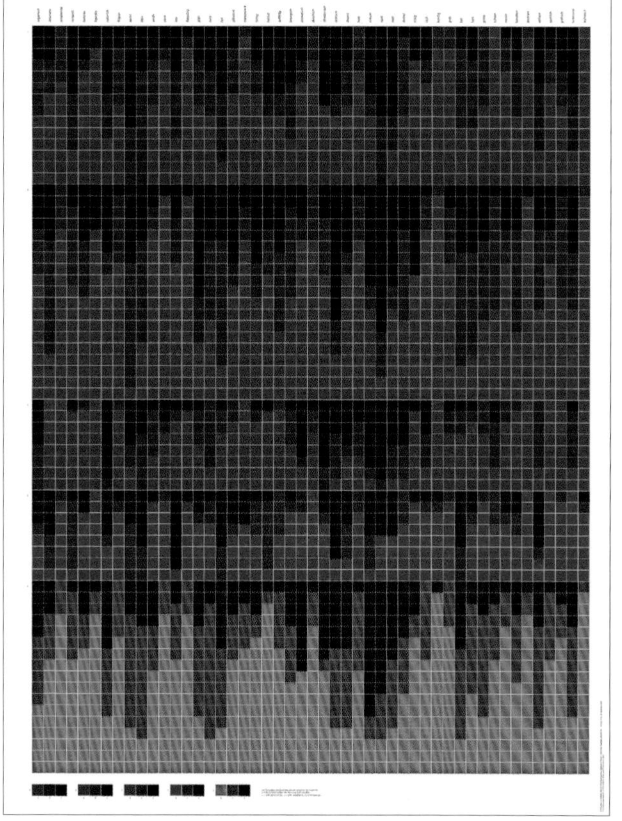
Women's Phone: Adjektivische Skalen II

04

02.1.6
可视化分析

设计师、统计学家、地理学者、建筑师以及通信专家的工作中都会涉及数据。《超越了饼状图和柱状图的数据可视化设计》表明，相比于传统的图解表示法，通过使用创新型可视化手段获得的数据记录可以提供高度描述性的信息。除完美的开发可视化外，该书籍还包含了在工作过程中所收集的所有数据。采用日本的书籍装订法，外页面包含原始资料（数据），而内页面则展示图形部分。书的内页打有小孔，以便能从这本书移除，将其展开后会显现出大型海报。

《女人的手机：正式的美学分析》说明的是将5个人建立的参考对象作为一种色谱。这一视觉转换的目的，是为了总结3项正式审美标准的典型特征（形状、颜色及表面结构）。

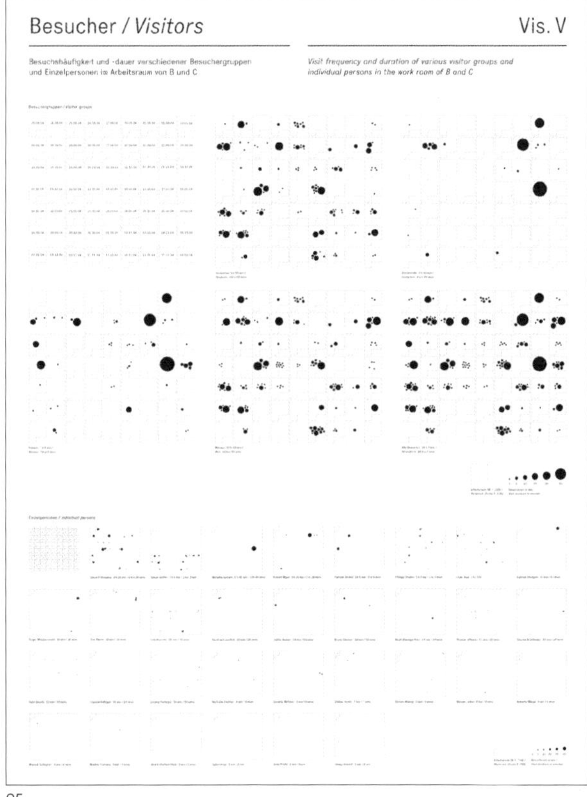

05

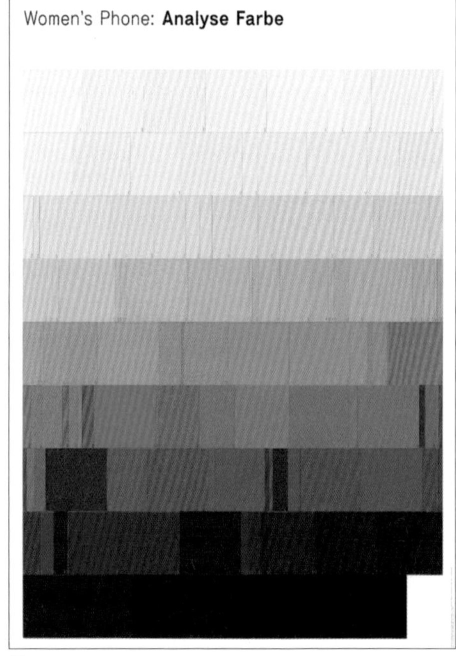

06

▸ 05

设计者：Von B und C
Barbara Hahn, Christine Zimmermann
作品名：*Data Visualization Beyond Pie*
克里斯托夫·梅里安出版社

▸ 06

设计者：Von B und C
Barbara Hahn, Christine Zimmermann
作品名：*Women' s Phone: Formal Aesthetic Analysis*
德国电信科研实验

▸ 07

设计者：Von B und C
Barbara Hahn, Christine Zimmermann
作品名：*Data Visualization Beyond Pie Charts*
图表和图形
克里斯托夫·梅里安出版社

Zusammenarbeit / *Collaboration*

Vis. IX

Gemeinsame und getrennte Arbeitsphasen von B und C, nach vier Arbeitsweisen unterschieden

Mutual and separate work phases of B and C, differentiated according to four work modes

Arbeitsphasen im Zeitverlauf / *Work phases over a period of time*

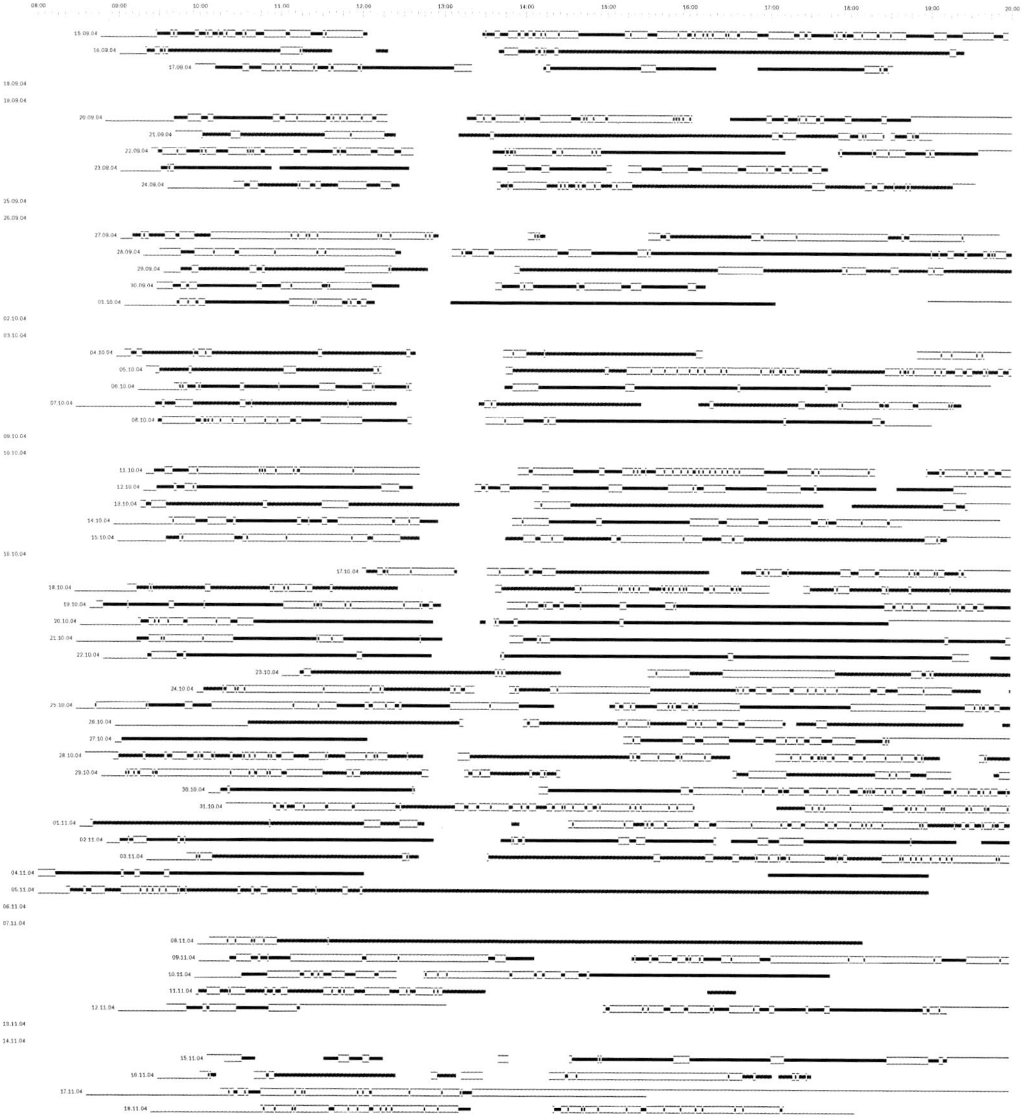

08:00 09:00 10:00 11:00 12:00 13:00 14:00 15:00 16:00 17:00 18:00 19:00 20:00

15.09.04
16.09.04
17.09.04
18.09.04
19.09.04
20.09.04
21.09.04
22.09.04
23.09.04
24.09.04
25.09.04
26.09.04
27.09.04
28.09.04
29.09.04
30.09.04
01.10.04
02.10.04
03.10.04
04.10.04
05.10.04
06.10.04
07.10.04
08.10.04
09.10.04
10.10.04
11.10.04
12.10.04
13.10.04
14.10.04
15.10.04
16.10.04
17.10.04
18.10.04
19.10.04
20.10.04
21.10.04
22.10.04
23.10.04
24.10.04
25.10.04
26.10.04
27.10.04
28.10.04
29.10.04
30.10.04
31.10.04
01.11.04
02.11.04
03.11.04
04.11.04
05.11.04
06.11.04
07.11.04
08.11.04
09.11.04
10.11.04
11.11.04
12.11.04
13.11.04
14.11.04
15.11.04
16.11.04
17.11.04
18.11.04

Aufteilung der gesamten Arbeitszeit / *Distribution of the entire working time*

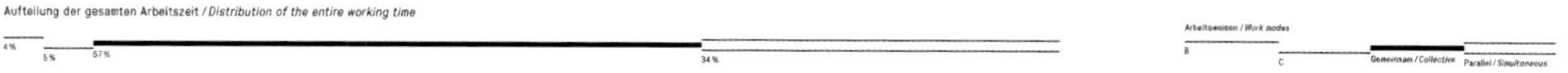

4 % 5 % 57 % 34 %

Arbeitsweisen / *Work modes*

B C Gemeinsam / *Collective* Parallel / *Simultaneous*

07

02.1.7
基于时间

“Unix”时间是一个描述时间点的系统，定义为自“UNIX”新纪元——1970年1月1日协调世界时间（UTC）午夜开始所经过的秒数。任何时间点都可以转变为其连续的总计秒数。在2009年2月13日（星期五）23时31分30秒时，“Unix”时间达到1234567890。“Unix”时间戳系列打印机为标志这一时刻并选取2009年中其他关键日程表而设计，如2009年1月20日巴拉克·奥巴马总统进行的就职典礼。

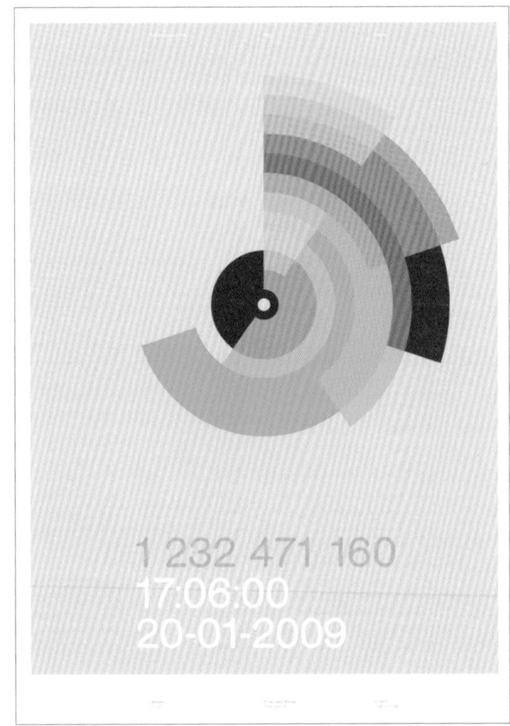

01

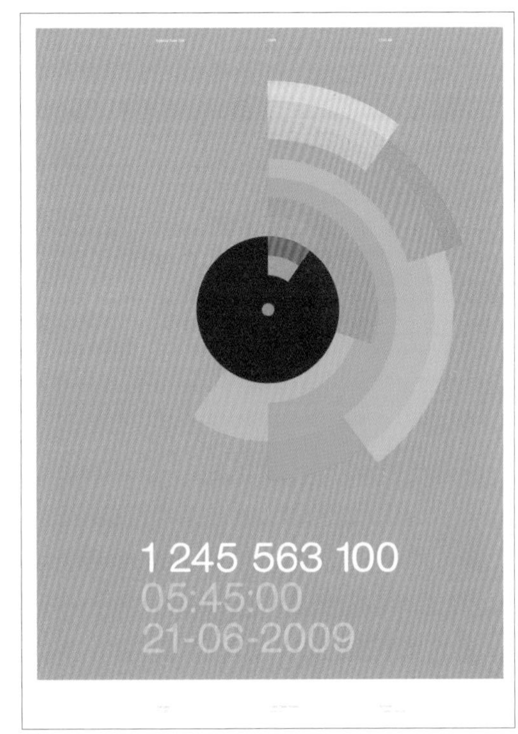

02

▸ 01

设计者：Jason Delahunty, Cody Delahunty
作品名：*The Inauguration of President Barack Obama*
海报

▸ 02

设计者：Jason Delahunty, Cody Delahunty
作品名：*Unix Time Stamp Solstice*
海报

▸ 03

设计者：Jason Delahunty, Cody Delahunty
作品名：*Unix Time Stamp 1 234 567 890*
海报

Friday Feburary 13th
2009
23:31:30
1 234 567 890
23:31:30
13-02-2009
Edition
1 of 25
Unix Time Stamp
Progression
Arrival
Graphic Design

03

02.1.8
社交混杂

伦敦市长基金的一则报告突出肖尔迪奇自治区的统计数据。这一报告的传播集中在该地区的社区，使用简单的人类形象表及充满生气的色彩来图解不同行业中的伦理交融。

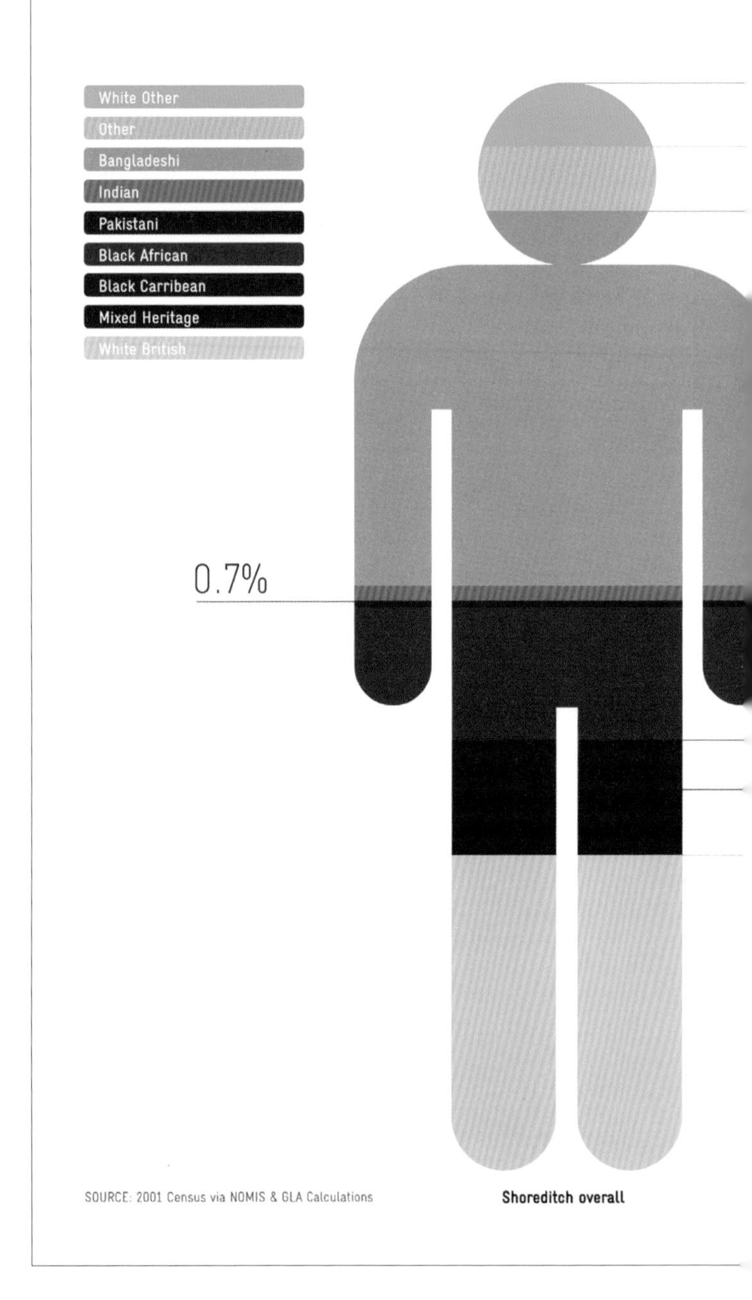

01
设计者：April
作品名：*London Mayor's Fund*
报告

Shoreditch has a wide range of communities from the Bangladeshi community of Bethnal Green, to the Turkish community of Hackney and the traditional white working class estates of Central Street and Finsbury. Other communities include the Afro-Caribbean, Kurdish and Somalian communities.

Of all children living in the area:

- 34% are from the Bangladeshi community, with a strong representation (up to 70%) in the eastern wards of Tower Hamlets
- 29% are white British, with a higher concentration in the Islington wards (55%)
- 17% are from the African-Caribbean community, with up to 35% in the Hackney wards
- 42% of children are Muslim, 35% Christian
- 14% of children are born outside the UK

Shoreditch is therefore a **'Community of Communities'**, and a one-size-fits-all approach to problem solving is not feasible. To deliver on our mission, we have resolved to work in partnership with the local Shoreditch communities. These comprise faith groups, residents associations, traders associations, schools, etc. This means that the local priorities and needs of the community are reflected in the programmes that we develop.

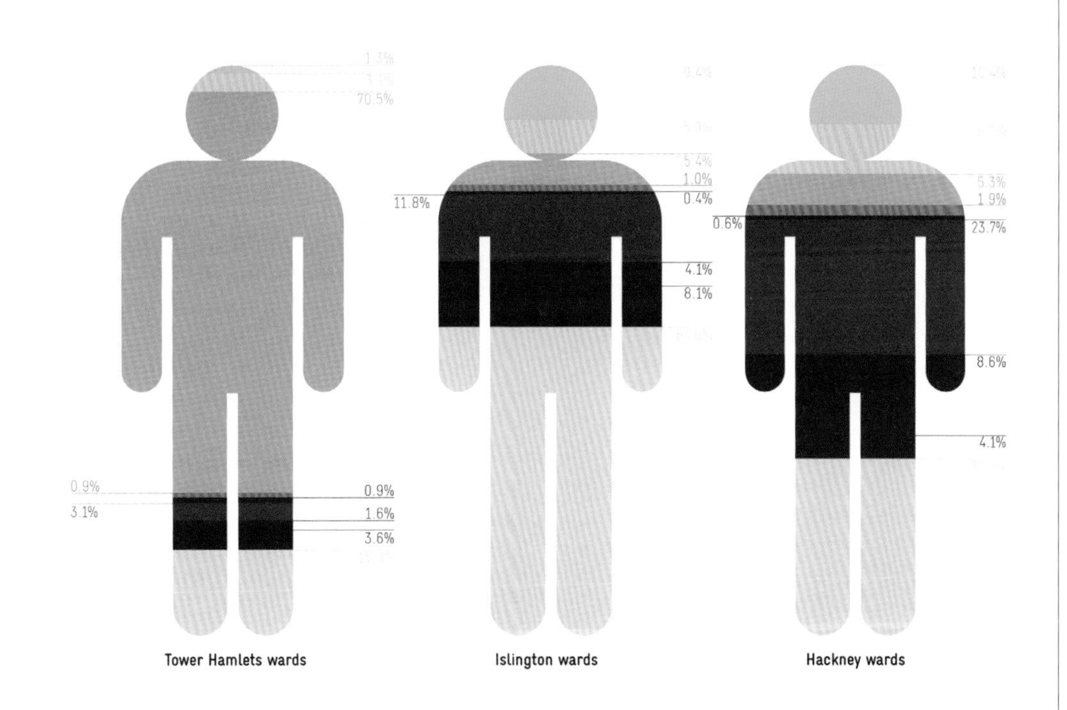

02.2 统计

02.2.1 触摸屏

《时代》杂志的iPad版受正在进行的一系列交互式图形支持。这里显示的是一系列描述英国北方和南方健康状况分布的信息图，其中一个信息图表说明的是自2000年起英国国债的巨额增长以及有关不列颠之战70周年纪念日的时间轴。

01

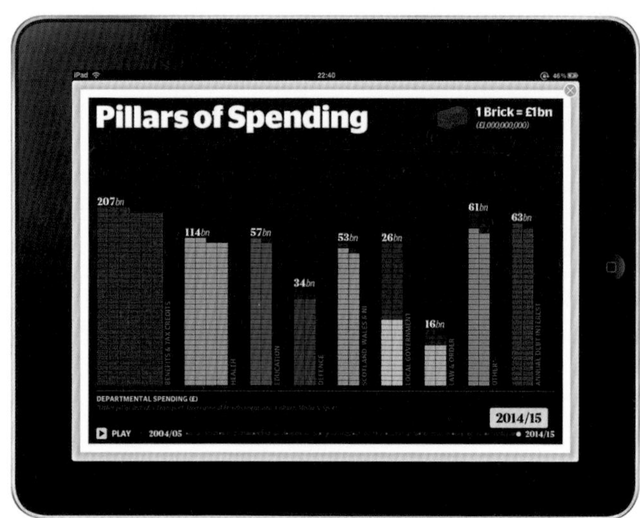

02

▸ 01

设计者：Applied Works
作品名：*The Times iPad edition*
不列颠之战时间轴

▸ 02

设计者：Applied Works
作品名：*The Times iPad edition*
消费额柱状图

▸ 03

设计者：Applied Works
作品名：*The Times iPad edition*
北方及南方健康状况分布

▸ 04
▸ 05

设计者：Applied Works
作品名：*The Times iPad edition*
债务墙

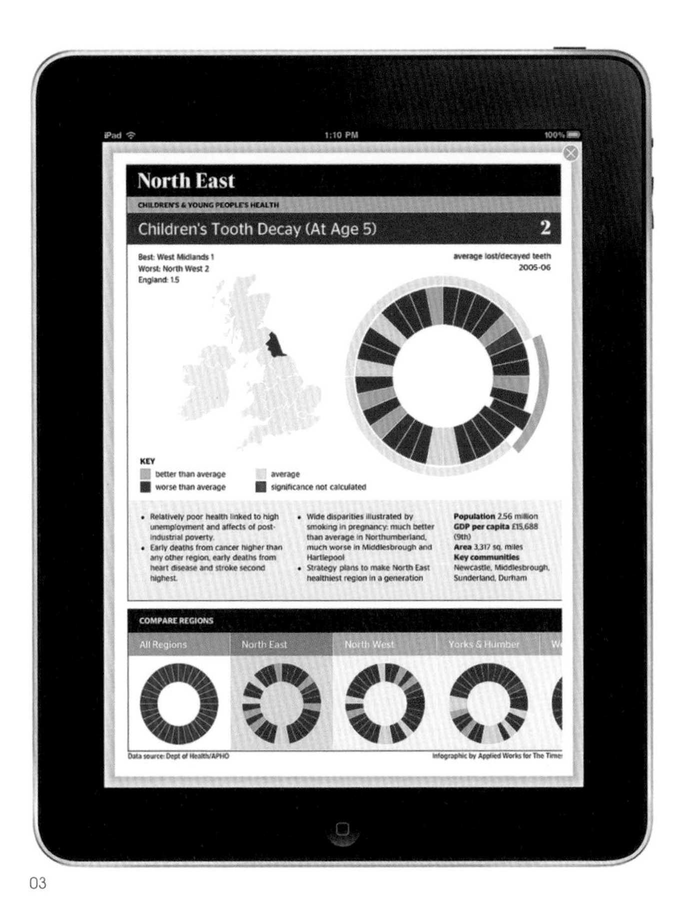
North East
CHILDREN'S & YOUNG PEOPLE'S HEALTH
Children's Tooth Decay (At Age 5)
2
Best: West Midlands 1
Worst: North West 2
England: 1.5
average lost/decayed teeth
2005-06
KEY
better than average
average
worse than average
significance not calculated
• Relatively poor health linked to high unemployment and affects of post-industrial poverty.
• Early deaths from cancer higher than any other region, early deaths from heart disease and stroke second highest.
• Wide disparities illustrated by smoking in pregnancy: much better than average in Northumberland, much worse in Middlesbrough and Hartlepool
• Strategy plans to make North East healthiest region in a generation
Population 2.56 million
GDP per capita £15,688 (9th)
Area 3,317 sq. miles
Key communities Newcastle, Middlesbrough, Sunderland, Durham
COMPARE REGIONS
All Regions
North East
North West
Yorks & Humber
Data source: Dept of Health/APHO

03

04

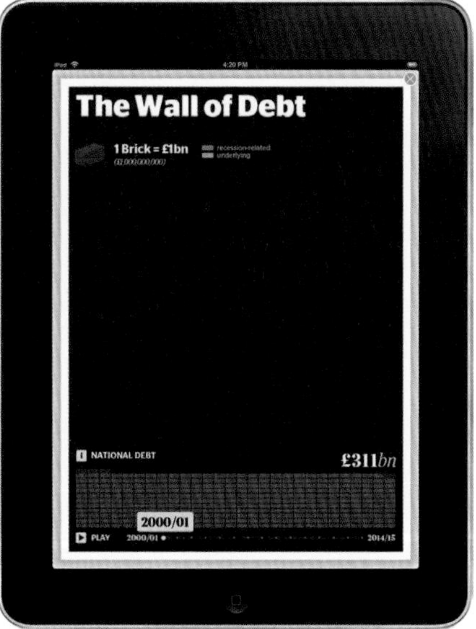
The Wall of Debt
1 Brick = £1bn
NATIONAL DEBT
£311bn
2000/01
PLAY
2000/01
2014/15

05

02.2.2
数据大爆炸

来自斯洛伐克文化部及塞尔维亚文化部的财务数据被直观化为可视数据信息项目。

《变形虫》是一个经典的柱状图，但其并不是在一根直轴上显示数值，而是所有栏目旋绕形成一个圆圈，通过峰值来表示数值。

《大爆炸》更多的是一种视觉感而非可视化数据的一项功能方法。随意摆放的每一圆圈均呈现出一个实际值，而中心圆或者称为爆炸源头区域的面积与所有白色圆圈区域组合起来的面积相等。

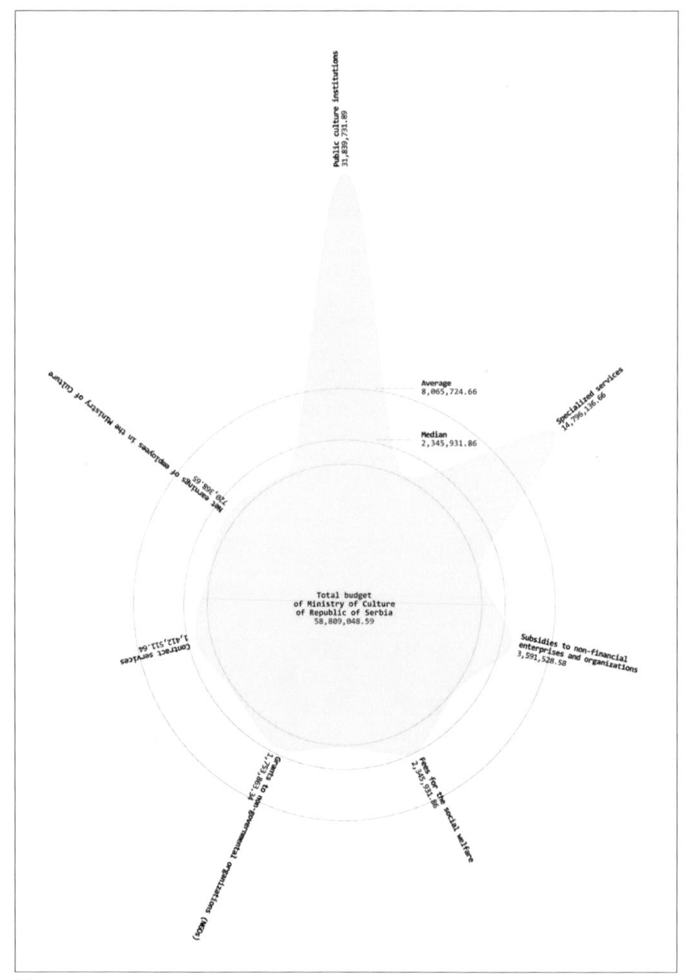

01

▸ 01

设计者：Ondrej Job
作品名：*Amoeba*
可视数据信息

▸ 02

设计者：Ondrej Job
作品名：*Big Bang*
可视数据信息

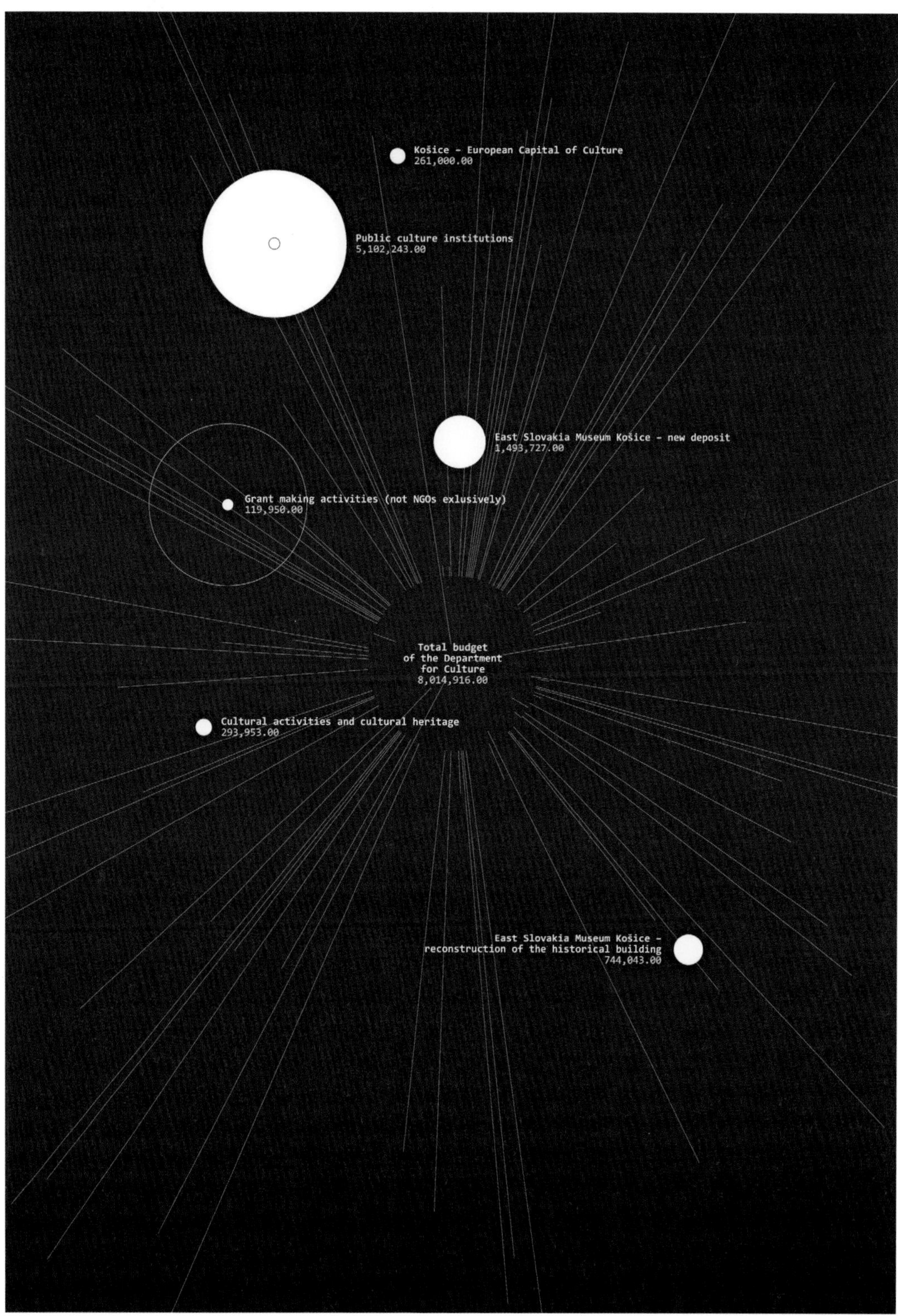
Košice – European Capital of Culture
261,000.00
Public culture institutions
5,102,243.00
East Slovakia Museum Košice – new deposit
1,493,727.00
Grant making activities (not NGOs exlusively)
119,950.00
Total budget
of the Department
for Culture
8,014,916.00
Cultural activities and cultural heritage
293,953.00
East Slovakia Museum Košice –
reconstruction of the historical building
744,043.00

02

02.2.3 尺度的概念

《每个人都曾在这个世界》是人类有历史记录以来在战争、大屠杀及种族灭绝中存活下来的人数对比被杀害人数的视觉再现。该可视化图像使用现有纸张区域和纸张损耗区域（模切圆）来分别呈现生与死的概念。人类自有文字可考的历史以来生存下来的人类总数约为776亿，在海报中表现为整个纸张面积（650 mm x 920 mm）。冲突中丧生的人数从史料书籍中整理得出，总数约为9.69亿。时序表跨越从公元前3200年至公元2009年——该时期横跨5 000年，见证了1 100余场战争。

图表左上角的点阵序列显示出过去5 000年里冲突数量的急剧增长（从左至右：公元前3 000年至公元2 000年）；最近1 000年达到暴力的顶峰时期。图表下部的大型阵点呈现的是接下来的1 000年以及所预测的人类冲突频率的惊人增长。

01

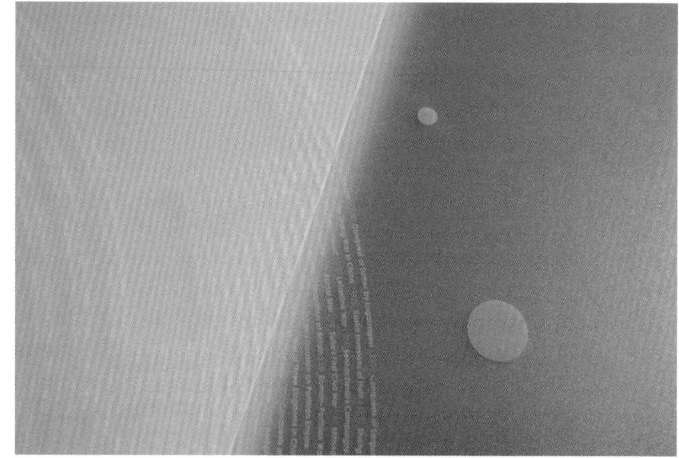

02

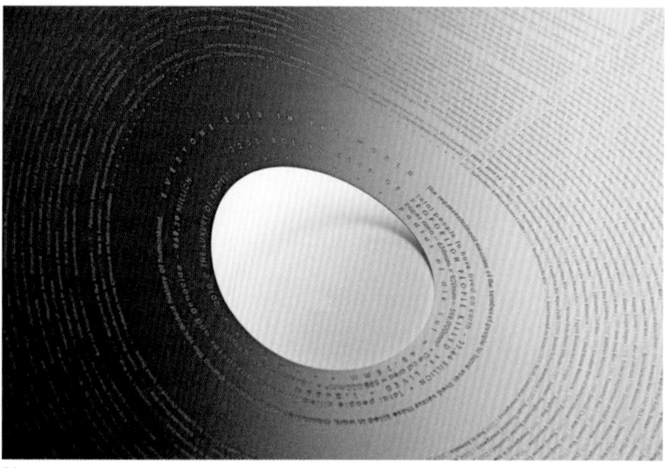

03

▸ 01
▸ 04

设计者：The Luxury of Protest
作品名：*Everyone Ever in the World (First Edition)*
具有缺口的大幅海报，在黑色塑料板上采用亮光透明油墨进行丝网印刷

▸ 02

设计者：The Luxury of Protest
作品名：*Everyone Ever in the World (Second Edition)*
具有缺口的大幅海报，在黑色塑料板上采用亮光乳白色油墨进行丝网印刷

▸ 03

设计者：The Luxury of Protest
作品名：*Everyone Ever in the World (third Edition)*
大幅海报，在棉纸上进行激光雕刻与激光切割

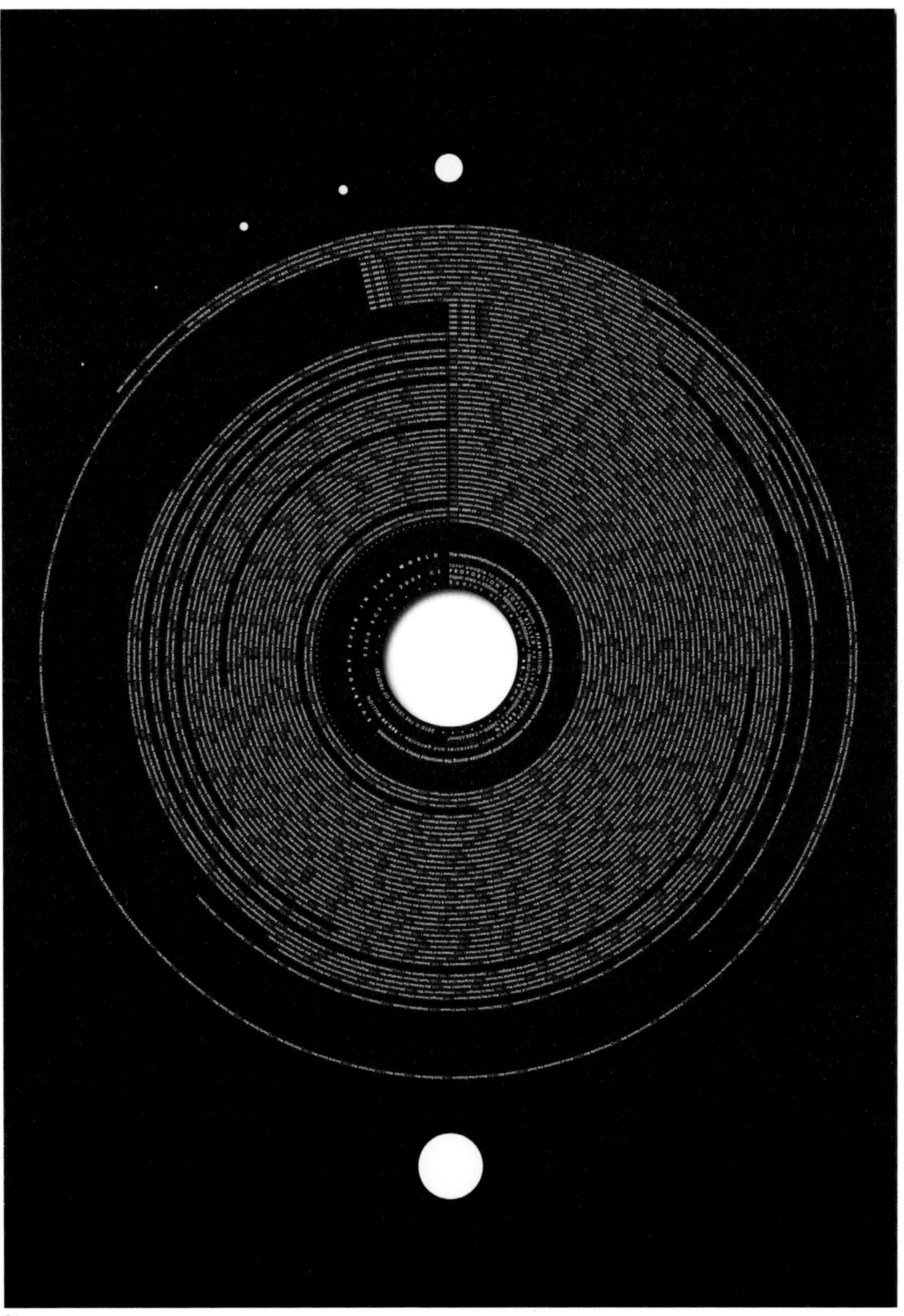

04

02.2.4
革命

《世界人权宣言》是一次图形的号令。该项目是由5位设计师发起的系列宣传海报组成，演说文章来自《世界人权宣言》，直接涉及抗议与言论自由的权利。基本的圆圈用来表达每篇文章的本质，通过单一的石墨铅笔点和标准黑点间的对比来呈现。石墨圆点表示每一副海报上的意见分歧，如《思想自由的权利……》这一文章。

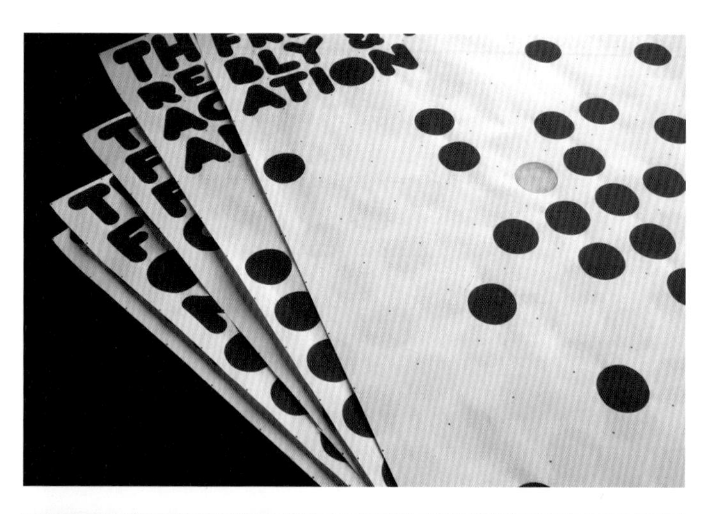

▶ 01
设计者：The Luxury of Protest
作品名：*The Universal Declaration*
系列海报

THE RIGHT TO FREEDOM FROM PERSE-CUTION

THE LUXURY OF PROTEST .COM

02.2.5
视觉分析

当《Questions》杂志发表可视化概念与设计时，正值瑞士宝盛银行举行第二届投资会议“为更好的世界而改变——绿色投资是可持续的吗？”说明性的可视化数据对有关能源消耗、水资源短缺、水质量或需求的现状加以描述，从而在视觉上完善了编辑部分。

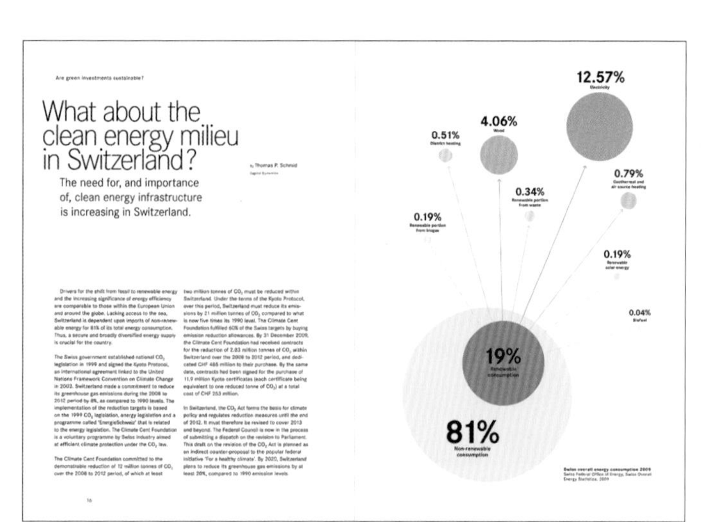

Are green investments sustainable?

What about the clean energy milieu in Switzerland?

The need for, and importance of, clean energy infrastructure is increasing in Switzerland.

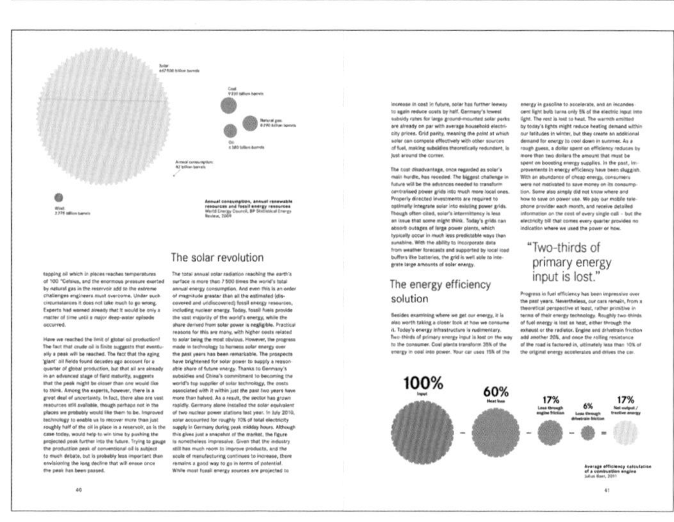

The solar revolution

The energy efficiency solution

“Two-thirds of primary energy input is lost.”

▶ 01

设计者：Von B und C
Barbara Hahn, Christine Zimmermann
作品名：*Questions*
瑞士宝盛银行

Are green investments sustainable?

Agriculture – how to meet tomorrow's demand

By Moritz Baumann
Julius Baer

Apart from cyclical swings and regional shortages, food prices, when adjusted for deflation, have declined over the past decades.

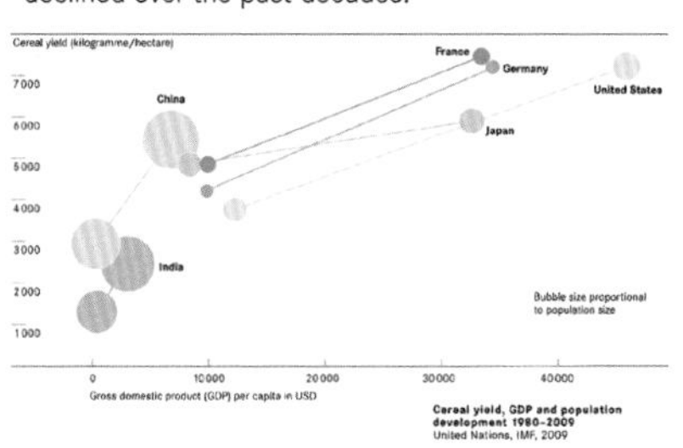

Cereal yield, GDP and population development 1980–2009
United Nations, IMF, 2009

42

Year

In million hectares

Global area of biotech crops
ISAAA, 2009

Based on that, one can surmise that food production has kept up with demand. Yet the world's appetite for agricultural goods is likely to grow. More mouths to feed due to an increasing global population and rising incomes in developing nations are leading to a change in diet favouring food that is higher in protein, and requires more energy to produce. The young and still-growing biofuels industry also has led to greater demand for agricultural products. Given that arable land is in tight supply in many parts of the world, the challenge for the future will be to boost productivity by growing more on every acre. The potential to achieve higher output must be assessed region by region: input availability, climate conditions and the state of development of the agricultural industry will all play a part. Nevertheless, opportunities exist that are already making a difference, and will continue to represent key advances in global agriculture to meet tomorrow's demand for food, feed and fuel.

Biotech

Massive gains in global agricultural productivity in the second half of the 20th century were largely due to the introduction of new seed varieties that ushered in the so-called Green Revolution. With big advantages offered by biotechnology, are we again witnessing the dawn of a new era in world agriculture? The answer is probably both yes and no. No doubt, genetically modified crops have proved to a large extent that they can increase yields and enhance food security. Such technology, however, faces hurdles with regards to gaining acceptance, amid concerns about its possible long-term impact on the environment and human health. Farmers' high dependency on the suppliers of these modified seeds is another matter of concern. What is often overlooked in such discussions are the potential environmental benefits to be had from biotech crops. How should the risks of biotech crops be assessed if

43

Are green investments sustainable?

they prove to markedly reduce the need for chemical fertilisers or pesticides, and allow less use of machinery and equipment, or perhaps require not as much irrigation? Given the challenges global agriculture faces to boost output in the decades ahead, and the fact that a few important food-producing countries, such as the United States, Brazil, India and China, already are increasingly using these new, modified crops, it appears likely that biotechnology will play a central role in the agricultural production of the 21st century.

"Over the past years, we have seen strong growth in fertiliser demand in developing regions, which is set to continue."

Fertilisers

The hunt for higher yields requires soils with the necessary nutrients for plant growth. To maintain productivity, nutrients taken out of the soil at harvest must be replenished. Fertilisers containing nitrogen, phosphorus and potash, as well as other minerals providing nutrients, are a key part of today's agricultural production. While in the developed world fertiliser use has remained fairly stable, such is clearly not the case in major developing producers like India and China where application of fertilisers still is below potential and not well balanced in terms of how the various ingredients are mixed. This makes it harder for farmers to achieve higher yields. Over the past years, we have seen strong growth in fertiliser demand in developing regions, which is set to continue. Fertilisers wrongly applied, however, pose significant hazards to the environment and to soil quality, and can also affect water resources, damage that might be irreversible. Such adverse outcomes have put severe

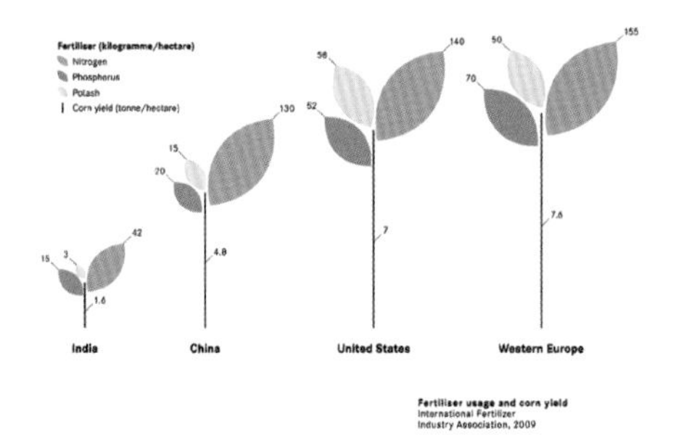

Fertiliser usage and corn yield
International Fertilizer
Industry Association, 2009

44

Year

Developing world – expected change in agricultural power mix
United Nations, 2003

strain on the ecosystems in the past. Yet environmental damage associated with fertilisers can be substantially reduced if these are applied correctly based on plant needs, while the benefits can be fully exploited. The challenge in the future will be to mitigate the potential damage to the environment that may result from fertilisers, even as the volumes in use are set to grow. Comprehensive solutions to manage nutrients include speciality fertilisers targeted to specific needs of plants and soil. Making professional advice available to farmers is also key. These aspects will play a major part in sustainable use of fertilisers in the future.

Agricultural equipment

In developed countries, mechanised production has largely replaced human labour in agriculture. This has facilitated large-scale farming and has allowed for enormous gains in productivity. While countries in the developing world are still in the process of introducing more machinery as they increasingly industrialise, a large share of the work in many of these places is still done by hand labour and animal strength. More widespread use of machinery to produce food crops in these countries is needed as urbanisation and jobs in other sectors attract rural dwellers away from the land. The trend towards substituting mechanisation for backbreaking human labour is driving demand for basic, small-scale equipment. This is true especially in parts of the developing world where farms are still relatively small, and thus unsuited for heavy machinery. Meanwhile in the developed world, changes underway include 'precision farming practices'. GPS-navigated machinery and measuring soil quality via satellite are some of the ways farming and IT are being combined to allow far more precise applications of pesticides, water and fertilisers. Besides potential to increase profitability of production and reduce volatility of crop yields, such techniques can reduce the adverse effects on the environment caused by agriculture.

45

02.2.6
数据丰富

20世纪90年代初期，荷兰住房部、空间规划部与环境部出版了空间规划第四份报告的副刊。这一政策性文件的荷兰语字母缩写为“Vinex”，其赢得了声誉，同时也获得了恶名。在其初期，整个荷兰建立起成千上万的住房，这就是著名的Vinex区域。有时得到称赞，有时得到诋毁，它们一直是争论的来源。《Vinex 地图集》首次对整个Vinex地区进行了深入的说明，在20世纪90年代中期，借助航拍，用规划、现场资料及近期现场图片对52个区域进行了描述。

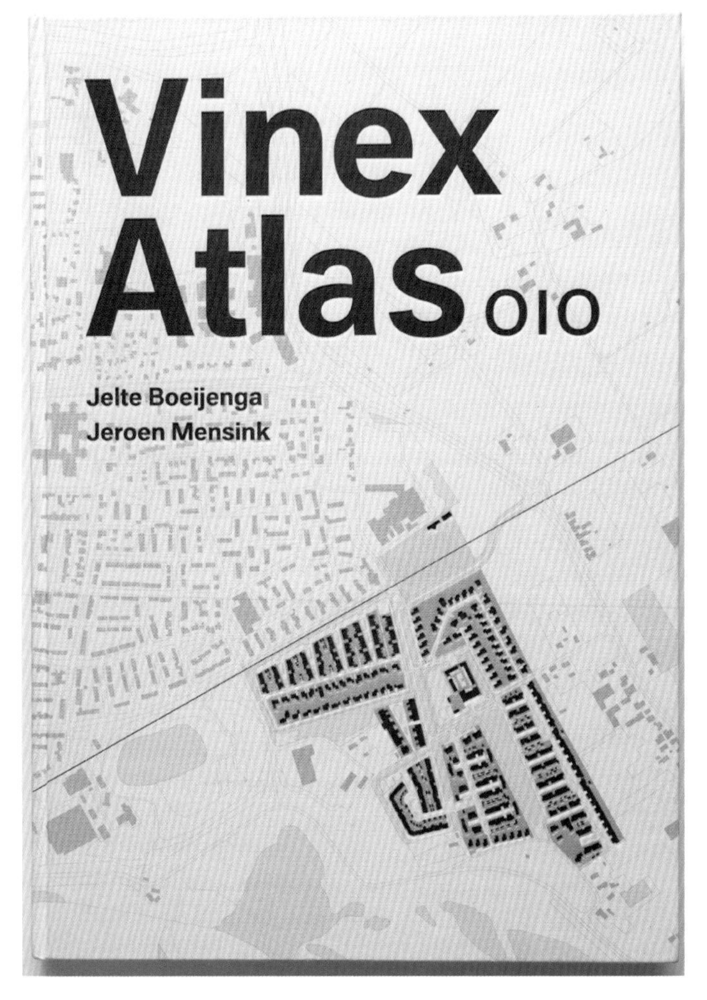

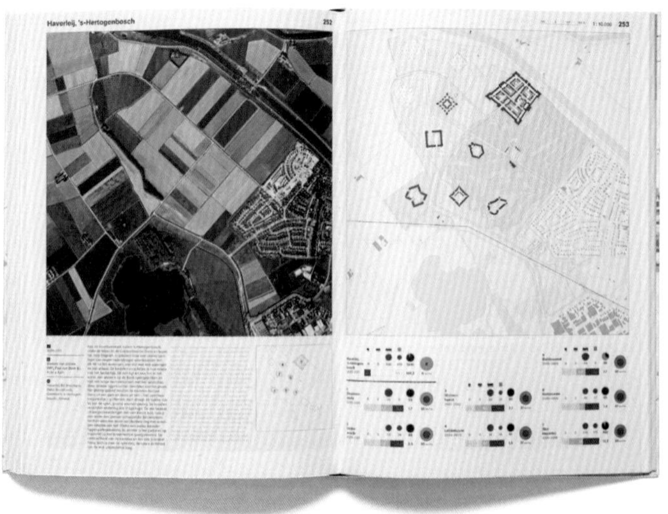

01
设计者：Joost Grootens
作品名：*Vinex Atlas*
010出版社

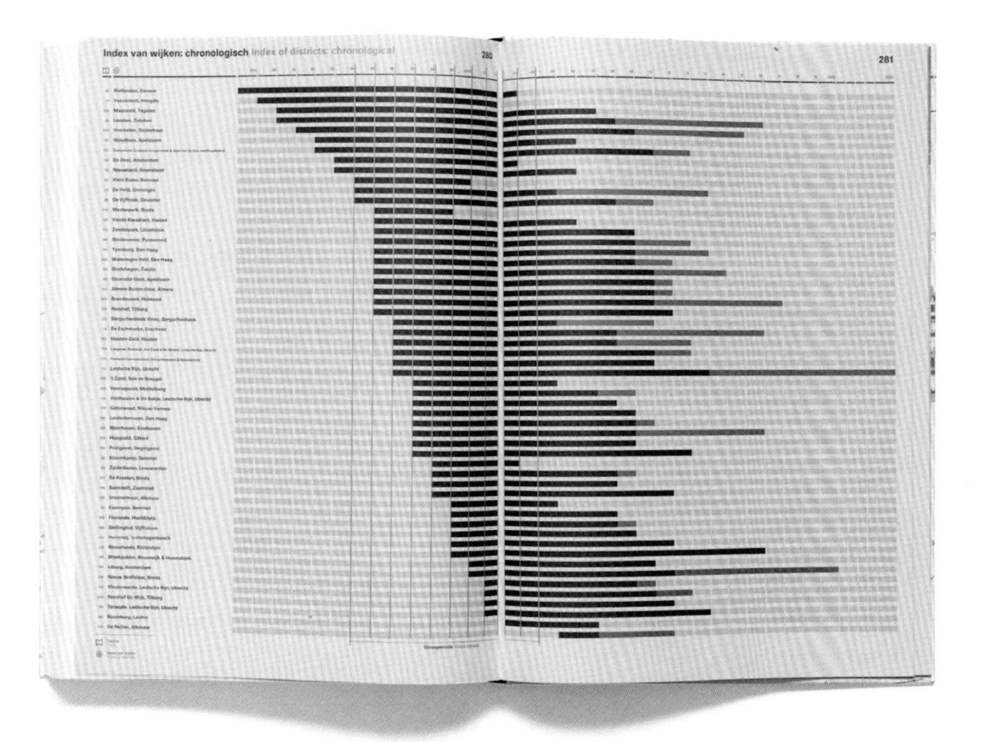
Index van wijken: chronologisch
280
281

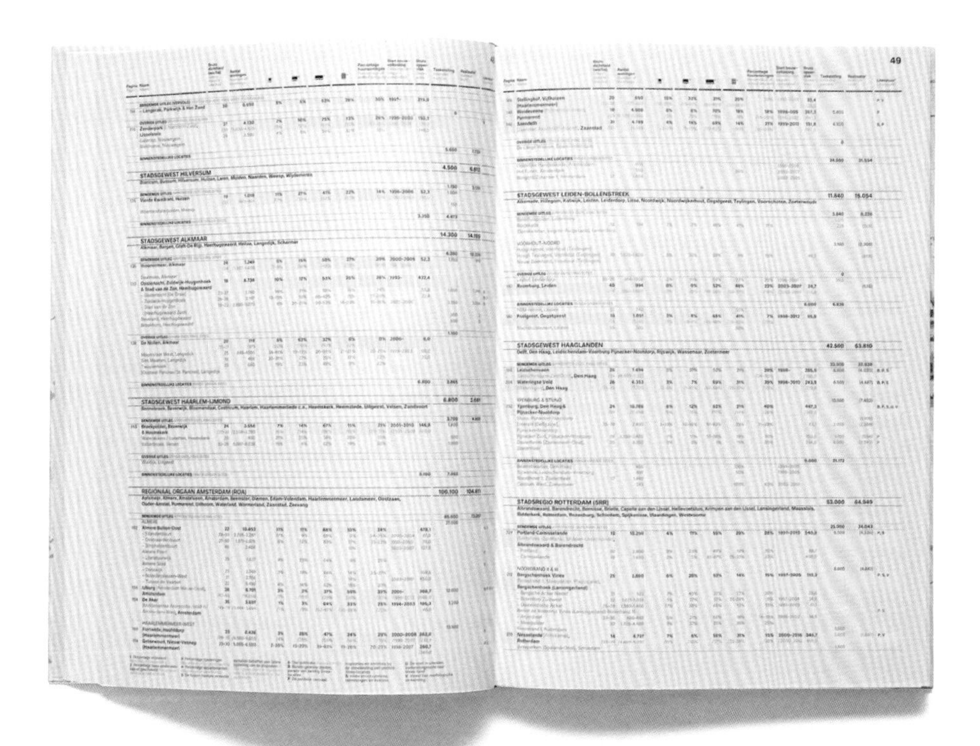
49
STADSGEWEST HILVERSUM
STADSGEWEST ALKMAAR
STADSGEWEST HAARLEM-IJMOND
STADSGEWEST LEIDEN-BOLLENSTREEK
STADSGEWEST HAAGLANDEN
STADSREGIO ROTTERDAM (SRR)

02.2.7
方向性

蒂宾根区议会大楼中的寻路标志系统采用天花板上的空间来设计方向性引导标示及统计数据，如到其他位置的距离等。

《英国平均降雨量》是从《黑暗的小时》与《光明的小时》海报（见第24~27页）而展开。海报的特点在于绘制有1971—2000年英国平均降雨量的图表。图表本身风格类似于倾盆的暴雨。相关条目直接绘制在与它们相关的气象站之上，形成了一幅英国的抽象地图。该地图采用黑色UV漆印于白色封底之上。

01

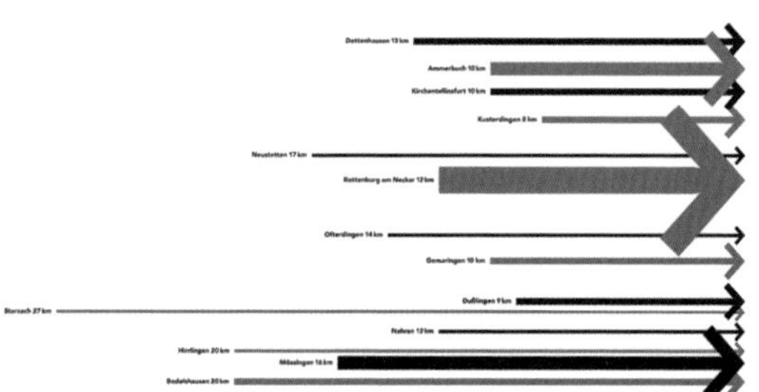

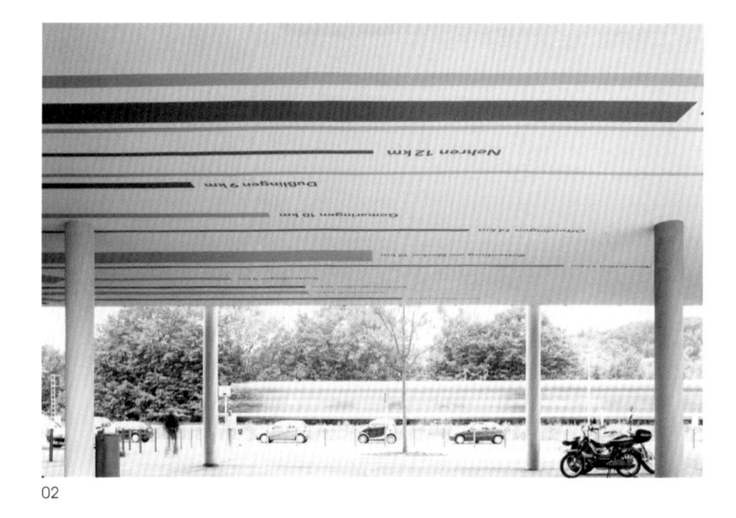

02

- 01
- 02
- 03
- 04

设计者：L2M3 Kommunikationsdesign
作品名：*TÜbingen District Council signage system*
导示系统

- 05

设计者：Accept and proceed
作品名：*Average UK Rainfall*
限量版海报

Bodelshausen 20 km
Hirrlingen 20 km

03

04

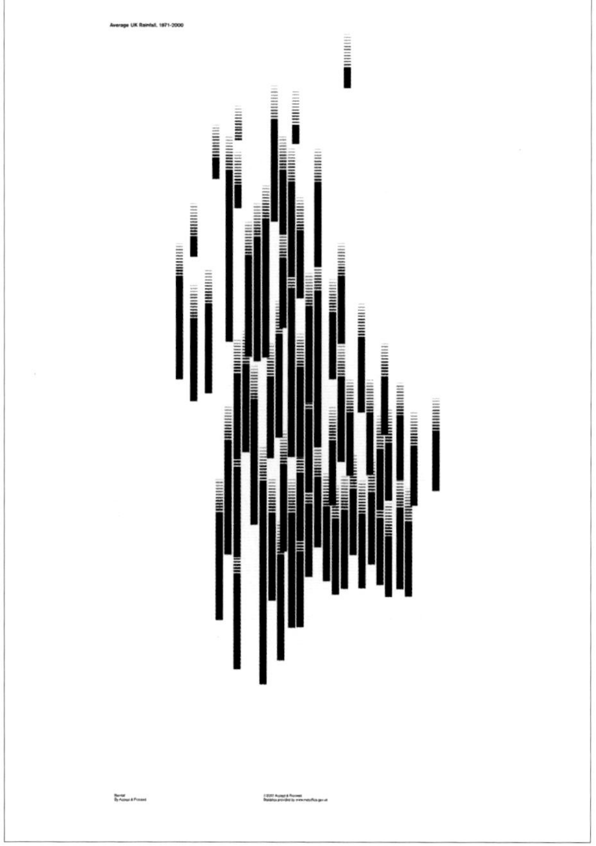
Average UK Rainfall, 1971-2000

05

02.2.8
工作流程

“本・桑德斯极地探险家”是一个统计数据丰富的网站，在2011年对“北3”的实时进度进行了跟踪，“北3”是一项由探险家及破记录长距离滑雪者进行的单独完成且无赞助的北极速度记录尝试。每小时一次的定位数据，每天更新的信息图表（显示纬度、每天的距离、雪橇质量以及温度）以及日常的博客文章使得人们能够实时了解桑德斯在为期36天的探险里的进度。

《工作日志》对设计师在某个5年期里的工作量进行了说明。工作量以相关年度里收到的电子邮件数目为基础。

《2007工作年》为一份日历，显示了设计者在2007年里的工作量。每一圆点表示一天——圆点越大，表示工作量越大。年份中的每一个月都有自己的颜色。使用的所有数据建立在每天收到的电子邮件数量的基础之上。

ARCTIC OCEAN
ARCTIC CIRCLE
85°35′
LATITUDE (°N)
12.5
DAILY DISTANCE (NAUTICAL MILES)
SLED WEIGHT (KG)
21.5
TEMPERATURE
-27°C
DAYS 1 2 3 4 5 6 7 8 9 10 11 12 13 14 15 16 17 18 19 20 21 22 23 24 25 26 28 29 30 31 32 33 34 35 36
LATITUDE (°N)
SLED WEIGHT (KGs)
DAILY DISTANCE (NAUTICAL MILES)
TEMPERATURE (°C)

01

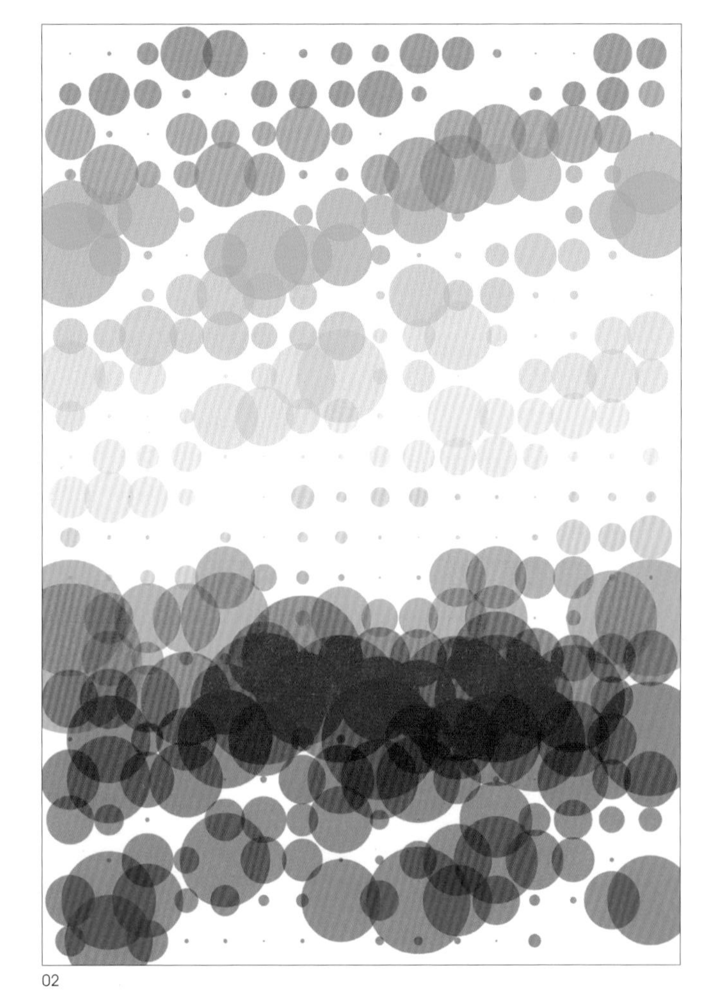

02

▸ 01

设计者：Applied Works
Studio8
作品名：*Ben Saunders Polar Explorer*
图形信息

▸ 02

设计者：Hörður Lárusson
作品名：*Work Calendar 04–09*
2004—2009年每年收到的电子邮件数量图

▸ 03
▸ 04

设计者：Hörður Lárusson
作品名：*Work Year 2007*
2007年中的每一天收到的电子邮件数量图

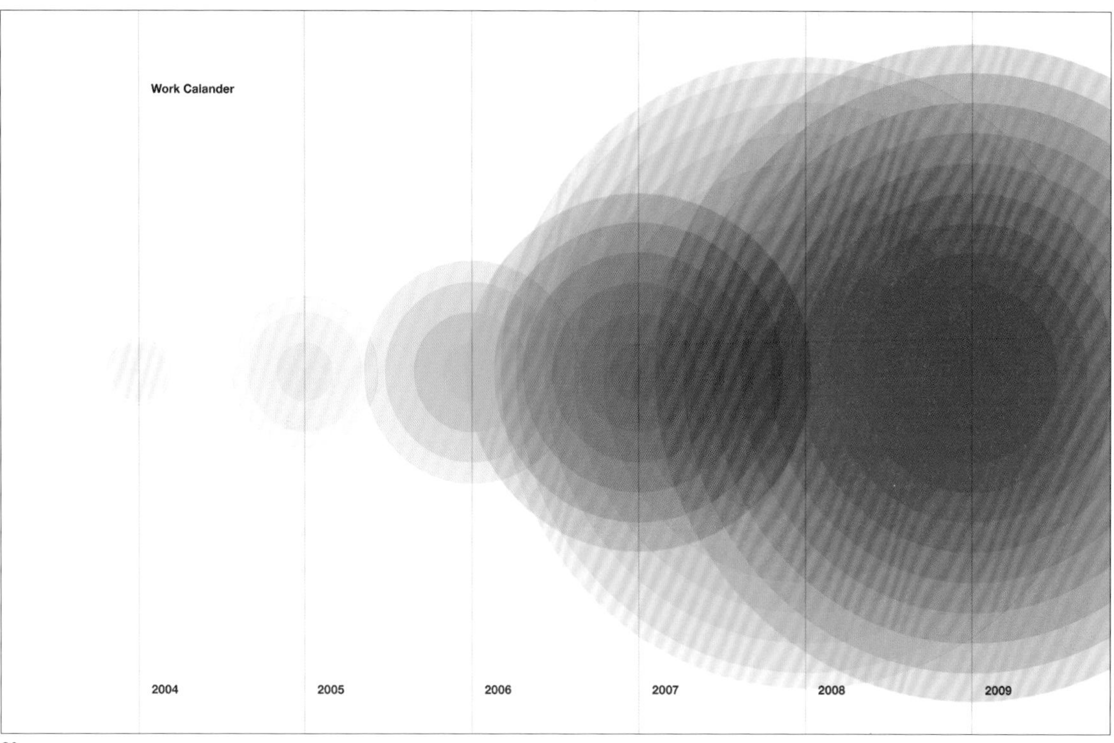

03

04

02.3
报告与说明

02.3.1
饼图

由“Form”设计的《媒介的信任》报道与说明如何设计一种简洁的解决方案来使说明信息最简化。此饼图采用单一的蓝色进行印制，在清晰的圆盘周围印有用词谨慎的注释。将大规模数字数据挑选出来放置于边缘，并通过添加颜色来突出显示此报道。

国际专业保险公司西斯考克斯（Hiscox）的年报保持色彩模式的最简化，依靠红色、灰色及黑色来呈现。同样地，图表也相当内敛和小巧。

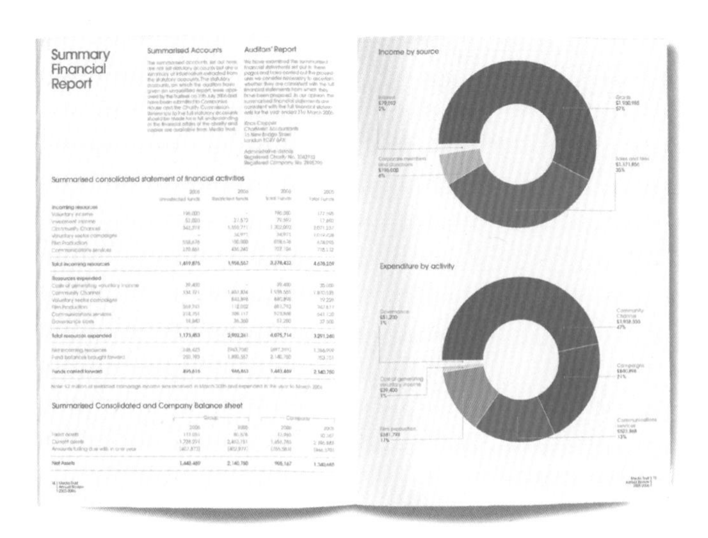

01

▸ 01

设计者：Form
作品名：*Media Trust*
报道与说明

▸ 02

设计者：Browns
作品名：*Hiscox 2010*
西斯考克斯保险公司2010年年度报告与报表

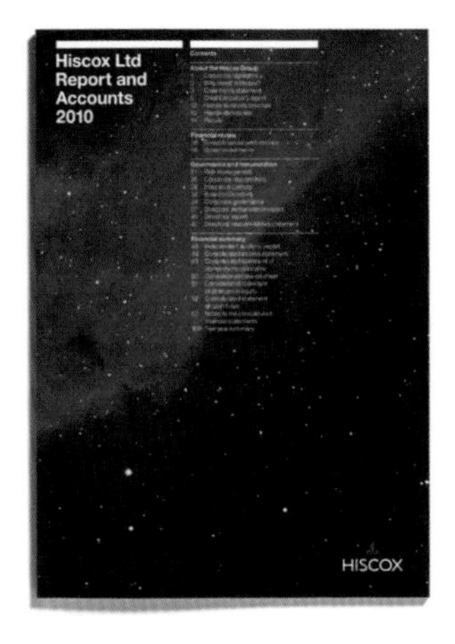
Hiscox Ltd
Report and
Accounts
2010
HISCOX

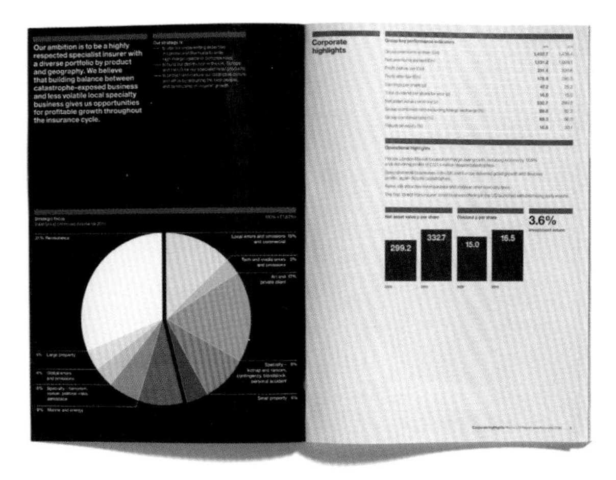
Our ambition is to be a highly respected specialist insurer with a diverse portfolio by product and geography. We believe that building balance between catastrophe-exposed business and less volatile local specialty business gives us opportunities for profitable growth throughout the insurance cycle.
Corporate highlights
3.6%
299.2
332.7
15.0
16.5

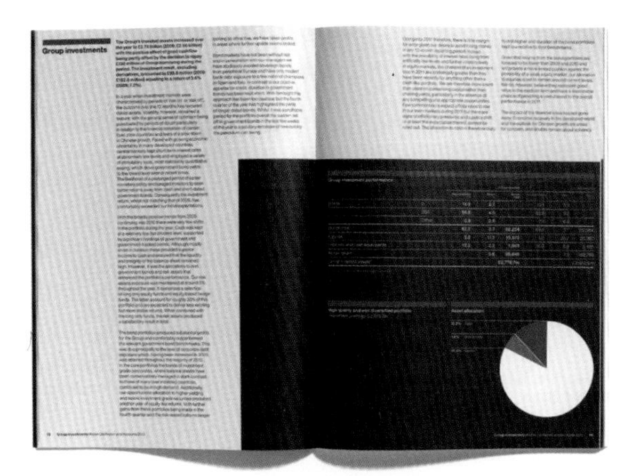
Group investments

Group investments continued
£2,779.7m
Risk management

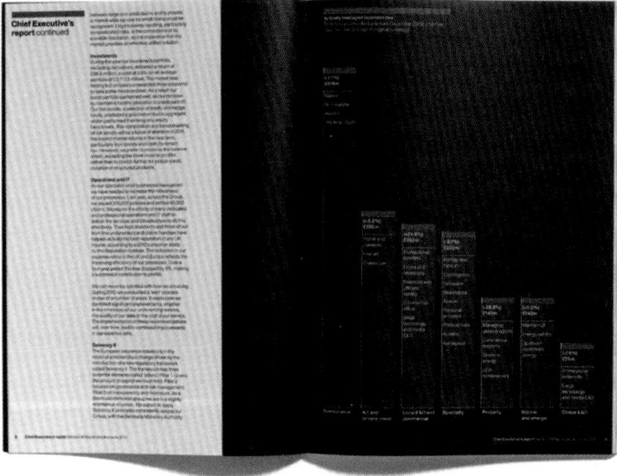
Chief Executive's report continued

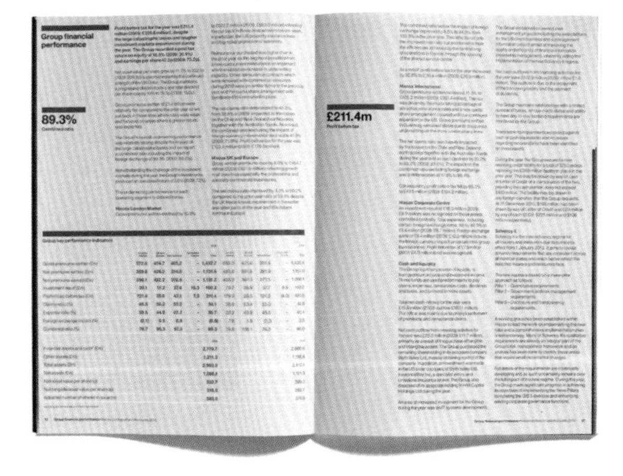
Group financial performance
89.3%
£211.4m

02

02.3.2 柱状图

英国国家设计委员会是英国针对设计而设立的国家级战略主体。由“Bibliotheque”设计的《2007/2008年度回顾》提出其政策的三大关键领域：竞争力、可持续性及创新。与传统的图示方法相反，其生动地表达了这些领域的增长。图形与图表漂浮在红色环境中（设计委员会公司的颜色），形成统计信息中时常缺乏的可表现情感和形象化的要素。

“Stapelberg & Fritz”为《E&A杂志》创作的信息图表设计，采用白色印制于黑色的廉价新闻用纸之上。原始数据保持简单，且统计信息去除所有的装饰性元素，用大型饼图及粗体大号数字占据设计的中心。

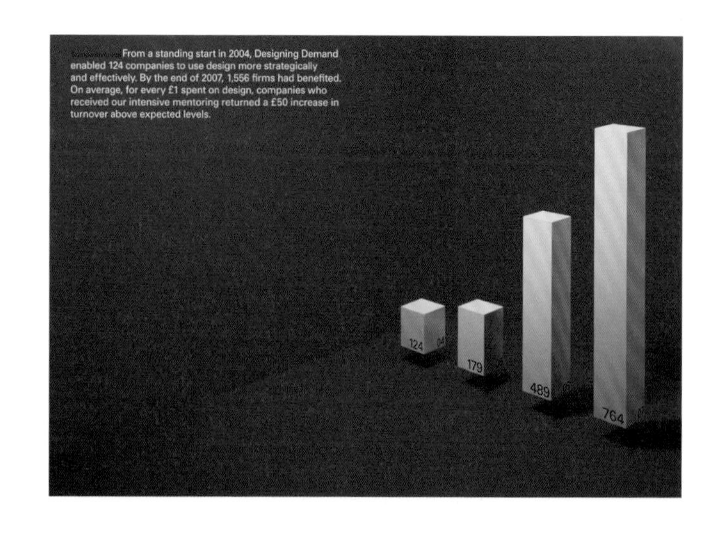

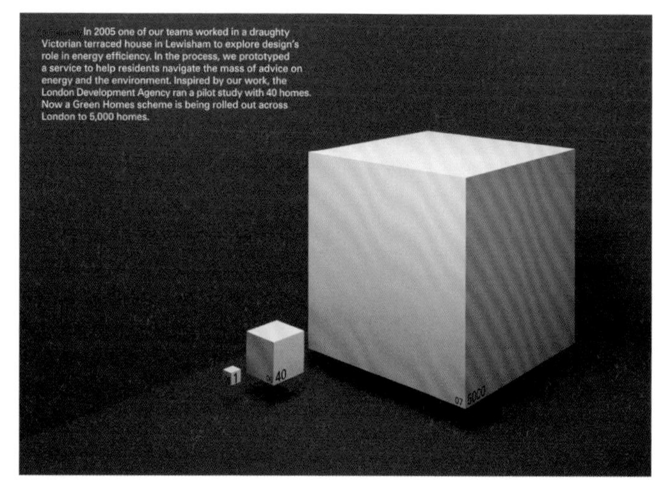

01

▸ 01

设计者：Bibliotheque
作品名：*07/08 Annual Review*
设计委员会

▸ 02

设计者：Stapeliberg & Fritz
作品名：*E&A Magazine*
信息图表设计

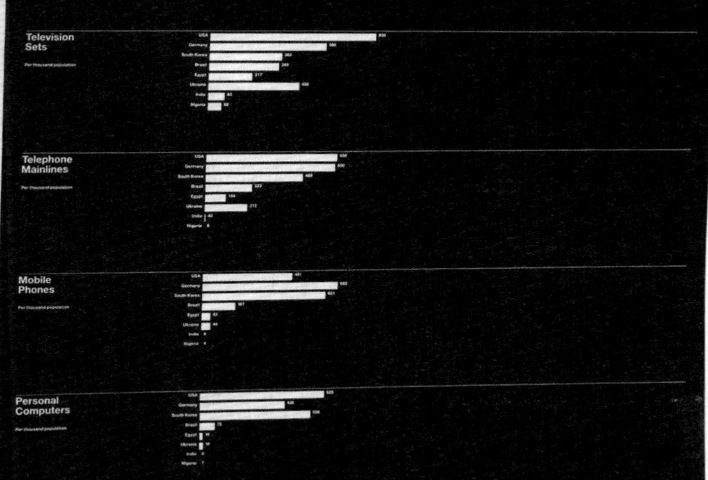

Table 2

Annual Expenditure on
Luxury Items Compared with
Funding Needed to
Meet Selected Basic Needs

In Billion Dollars

	18		12
	17		19
	15		5
	14		10
	11		1.3

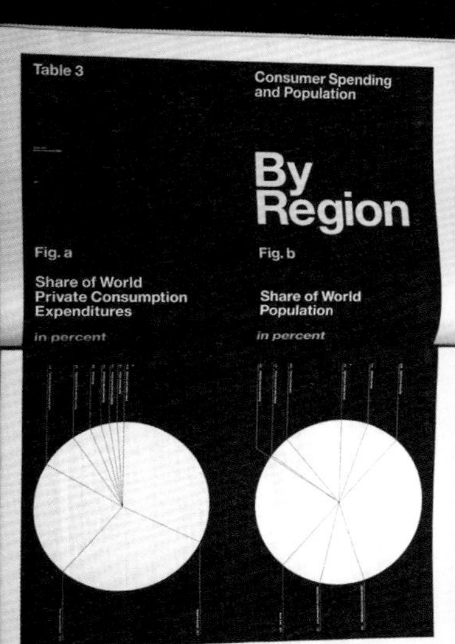

02.3.3
无图表

日本医药公司“大冢制药”1979年的年报中所有饼状图、图表及其他图示设置都被去掉，使数据远离了枯燥无味。销售数据通过比较量表中相互间百分比的重现来说明，每部分采用不同的颜色来表示。该公司之后一份报告的封面在2003年度的两个0上下功夫，将其作为一种图形工具来显示公司运营中的细节。

由商业艺术公司为《活跃的社区网络》设计的《打破障碍的报告与摘要》由72页的学术报告组成，报告大部分由列表资料及编号部分组成。这种数字编号系统形成了报告的主干，因为其他的直观材料很有限。

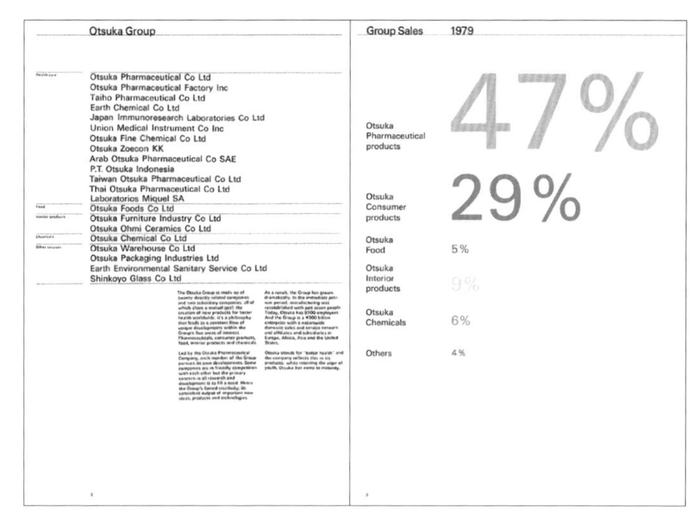

01

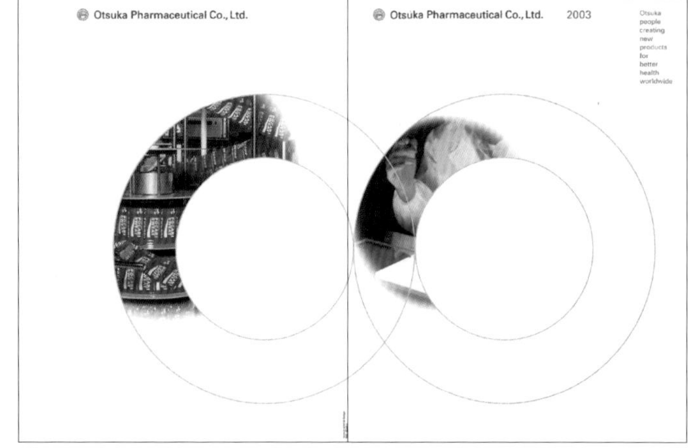

02

▶ 01

设计者：Helmut Schmid
作品名：*Otsuka Pharmaceutical 1979*
摘自该公司年度报告和说明

▶ 02

设计者：Helmut Schmid
作品名：*Otsuka Pharmaceutical 2003*
年度报告及说明封面

▶ 03

商业美术
作品名：*Breaking Barriers Report and Summary*
活跃的社区网络

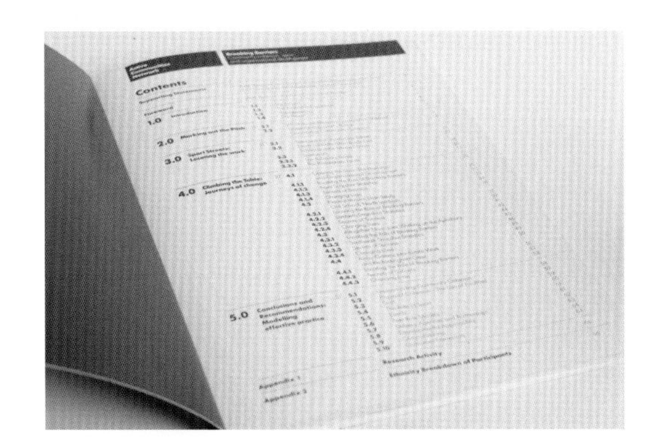
Contents

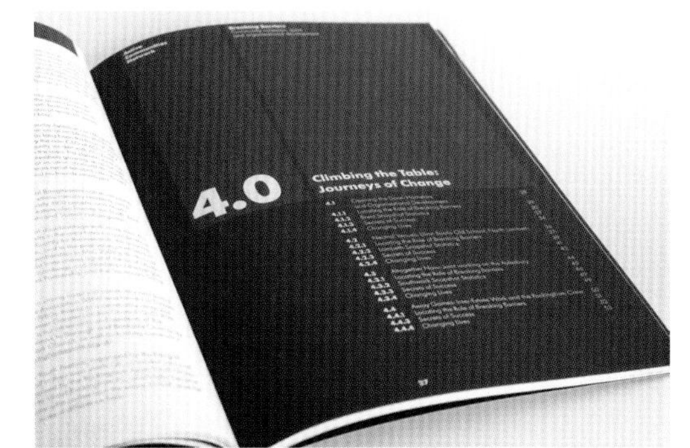
4.0
Climbing the Table:
Journeys of Change

4.3.3
Secrets of Success
4.3.4
Changing Lives

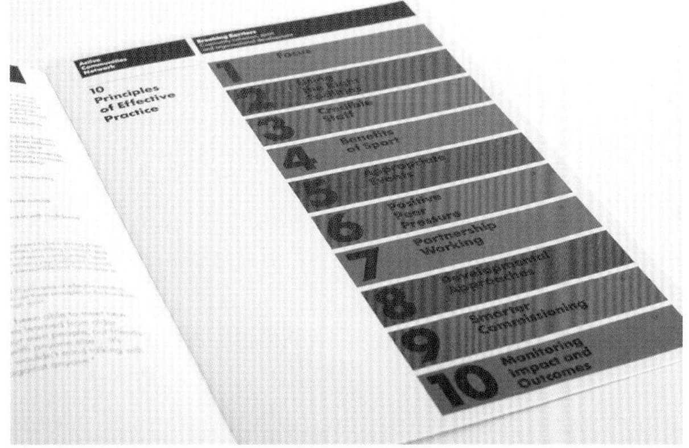
10 Principles of Effective Practice
1 Focus
3 Credible Staff
4 Benefits of Sport
5 Appropriate Events
6 Positive Peer Pressure
7 Partnership Working
8 Developmental Approaches
9 Smarter Commissioning
10 Monitoring Impact and Outcomes

10 Principles of Effective Practice
5 Appropriate Events

03

6.

5.11.2010
ab 21.30 h

Distribution-Lounge mit audiovisueller Performance
»Improvisations – Requiem for a Tree« von Robert Seidel (Weimar) und Michael Fakesch (Ex-Funkstörung, Rosenheim)

6.11.2010

Kunststiftung Baden-Württemberg

Vorträge
10–12.30 h

Wolfgang Knauff
Geschäftsführer Stiftung Kulturserver und Onlinefilm AG, Aachen Berlin: »Films Are Made to be Seen« – Distribution von Video im Internet am Beispiel von Onlinefilm.org

Stefanie Reis und Birgit Glombitza
KurzFilmAgentur Hamburg: Recht und billig? Die kreative Marke Kurzfilm – Erlösmodelle und Absatznischen im Weltvertrieb der KurzFilmAgentur Hamburg

Vorträge
14–17 h

Dr. Sabine Maria Schmidt
Kunstmuseum Folkwang Essen: Videokunst im Museum – ein Balanceakt zwischen technischer Entwicklung und dauerhafter Konservierung, Ausstellungsalltag und Forschung

Ivo Wessel
Kunstsammler und Geschäftsführer von iCodCompany, Berlin: Sammeln von Videokunst – Aspekte, Herausforderungen, Grenzen

Prof. em. Birgit Hein
Künstlerin Berlin: Kuratieren als Erweiterung der künstlerischen Arbeit

3./4.

3./4.11.2010
jeweils 10–17 h

Freie Kunstakademie Nürtingen e.V.

Workshop

Audiovisuelle Performance

Referenten

Robert Seidel und Michael Fakesch
Anhand von moderner Sound-, Animations- und Videosoftware stellen die beiden Künstler die Möglichkeiten vor, in Echtzeit korrespondierende Visuals und Soundarbeiten zu entwickeln.
Robert Seidel hat an der Bauhaus Universität Weimar studiert und diverse preisgekrönte abstrakte Animationsfilme sowie Visuals für Bands wie Passion Pit realisiert. Michael Fakesch war Mitglied des Elektroprojekts Funkstörung. Er produzierte Remixe u.a. für Björk und Sound Design u.a. für Vodafone, BMW und Philips.

Anmeldung bitte per e-mail an der FKN FKNuertingen@aol.com oder per Telefon 07022 533 00

Die Teilnahmegebühr beträgt 60,00 Euro. Die Teilnehmerzahl ist auf maximal 12 Personen begrenzt. Studenten der Freien Kunstakademie Nürtingen können kostenlos am Workshop teilnehmen.

DISTRIBUTE!

5./6.

DISTRIBUTE! 2
2010

5.11.2010
13–17 h

Merz Akademie Hochschule für Gestaltung Stuttgart Staatlich anerkannte Fachhochschule

Workhop

Strategien der Online-vermarktung für Künstlerinnen und Künstler

Referent

Wolfgang Knauff
Geschäftsführer Stiftung Kulturserver und Onlinefilm AG

Anmeldung bitte p
beim K
oder über die H
des Ku

2

03.
顺序

03.1　时刻表

03.1.1　简单与明了　090
03.1.2　地方的耐用性　092
03.1.3　整合系统　094
03.1.4　多重项目　096
03.1.5　线型项目　098
03.1.6　简洁性与复杂性　100

03.2　列表

03.2.1　连续时间　102
03.2.2　日期与时间　104
03.2.3　垂向层序　106
03.2.4　日期与数据　108
03.2.5　线与边框　110
03.2.6　超大号日期　112
03.2.7　分片　114

03.1
时刻表

03.1.1
简单与明了

“Image Now”设计工作室承担了都柏林公交车形象的总体重新设计，该项目扩展至所有线路图和时刻表。规划的方式在于为该公司的品牌注入新的生命，并且始终以简洁明了的方式诠释它。

从车的外观到街道设施，公交线的每一个元素都得以重新设计。路线图和道路引导通过删除多余的信息和视觉上的混乱，清晰性和明了性得以加强。例如时刻表的设计方法采用的是明确的公共汽车时间表和大号字体的线路编号。

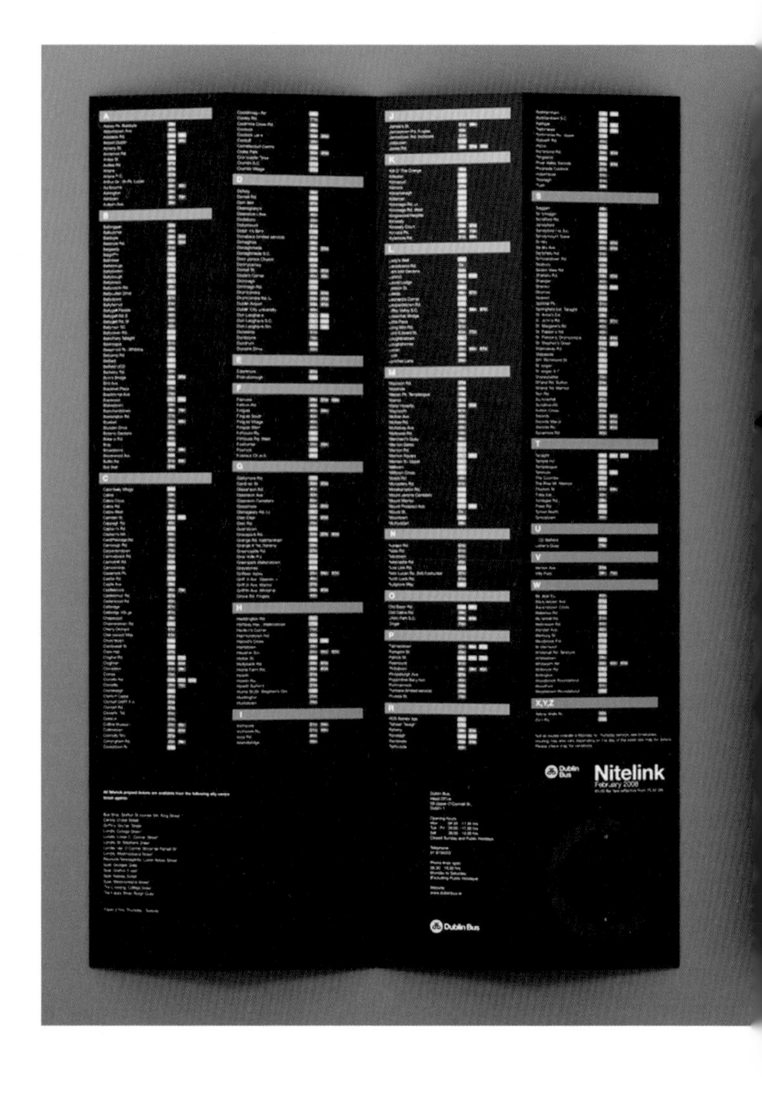

▸ 01
设计者：Image Now
作品名：*Dublin Bus*
新标识、线路图与时刻表

Westbound
Southbound

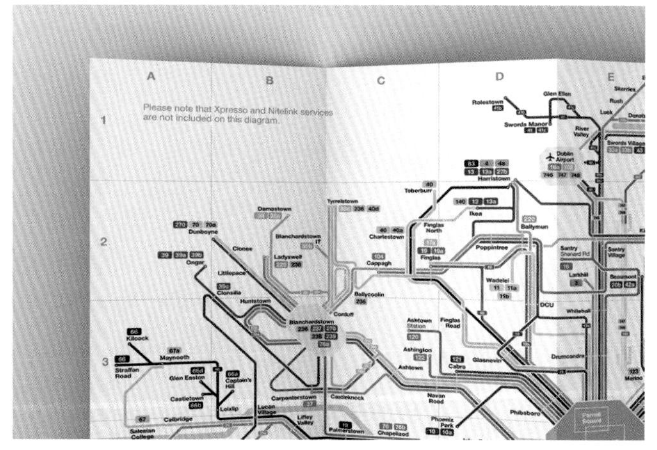

A
B
C
D
E
Please note that Xpresso and Nitelink services are not included on this diagram.

03.1.2
地方的耐用性

专门为冰岛西部城市伊萨菲厄泽设计的公交系统，是该市与周边城镇共用的公交系统，三辆公交车行使五条线路。简称为“全尺寸”系统，也反映了城镇轻松的氛围。

设计师对该地区的地图、时刻表及附加区域性信息进行了设计，并已在大约30个公交站内设立。

时刻表每年更换两次（冬季和夏季）。每个公交站的时刻表使用伊萨菲厄泽市图片的某一部分作为特色。想要见到组成这一图像的所有部分，游客们必须走遍每一站。

时刻表采用一种清晰简单的方式，使用字体最小值等宽字体版本。地名沿着时刻表顶部倾斜30度角，这有助于容纳该地区内某些冗长的地名。网格中的每一路线都具有不同的颜色，置于地名下方的小色板用于指明线路中的十字路口。

标志由西吉・艾格特森设计，金属框架由Stefan Petur Solveigarson设计。框架的设计与建造旨在抵御该地区冬天的极端天气。

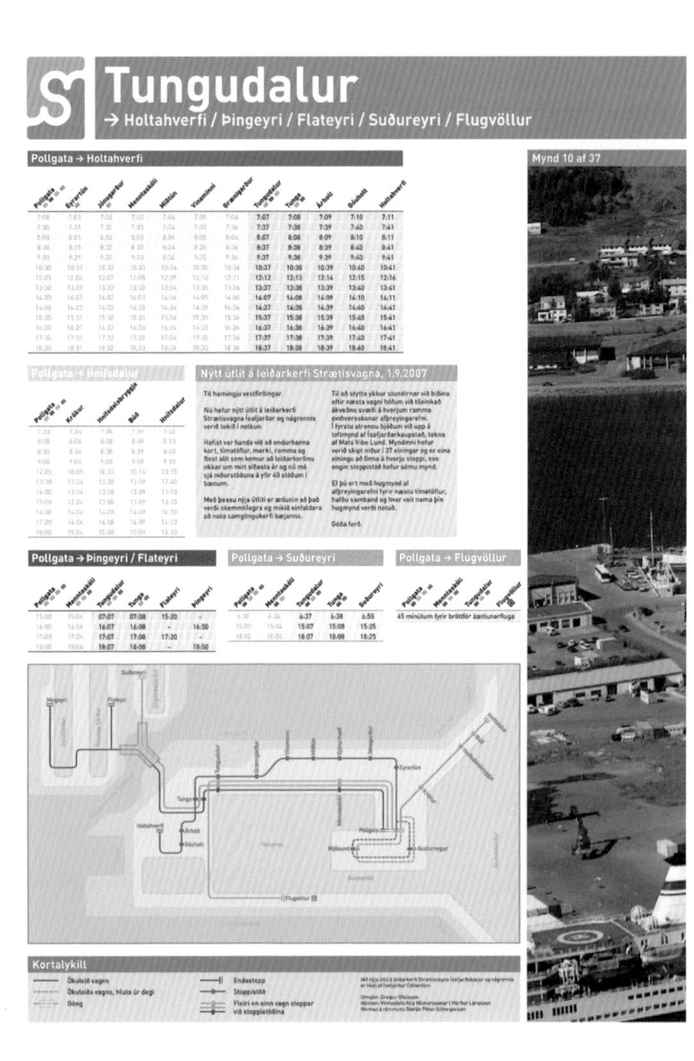

▸ 01

设计者：Atelier Atli Hilmarsson

作品名：*Public Transport of Isafjorour*

新型标志、路线图及时刻表

Góuholt

Hnífsdalur
50

03.1.3
整合系统

《谢菲尔德的连接》的理念在于将用户作为设计过程的中心，确保各部位的相关信息都是精心策划的。标志系统包括谢菲尔德城市公交及电车线路的局部街道图。公交与电车网线路示意图及时刻表遵循同类设计的原则，在视觉上相互统一。使用深浅不一的灰色来为彩色地图提供一种中性背景。

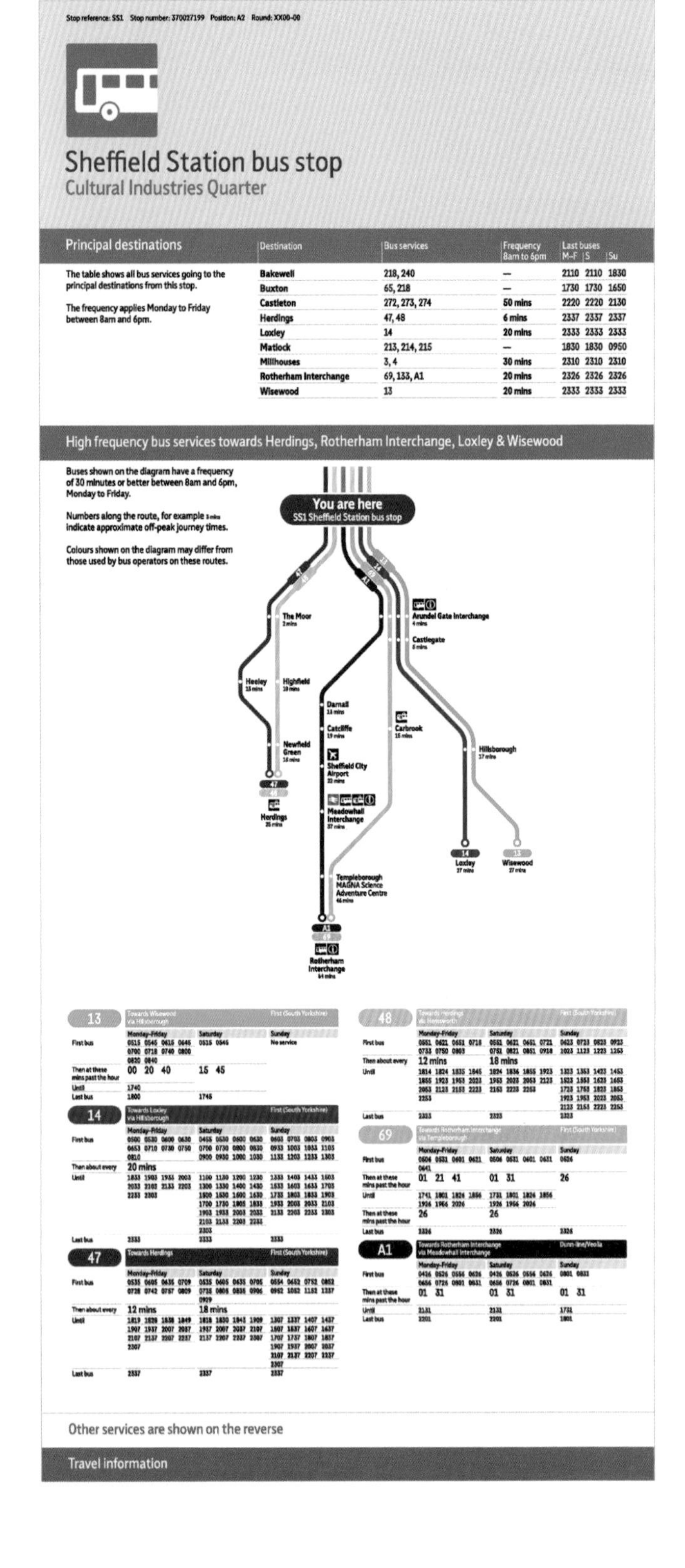

▶ 01

设计者：City ID, Atelier Works, Pearson Lloyd

作品名：*Connect Sheffield*

导视系统

T
Castle Square Supertram stop
Towards Meadowhall Interchange, Halfway & Herdings Park

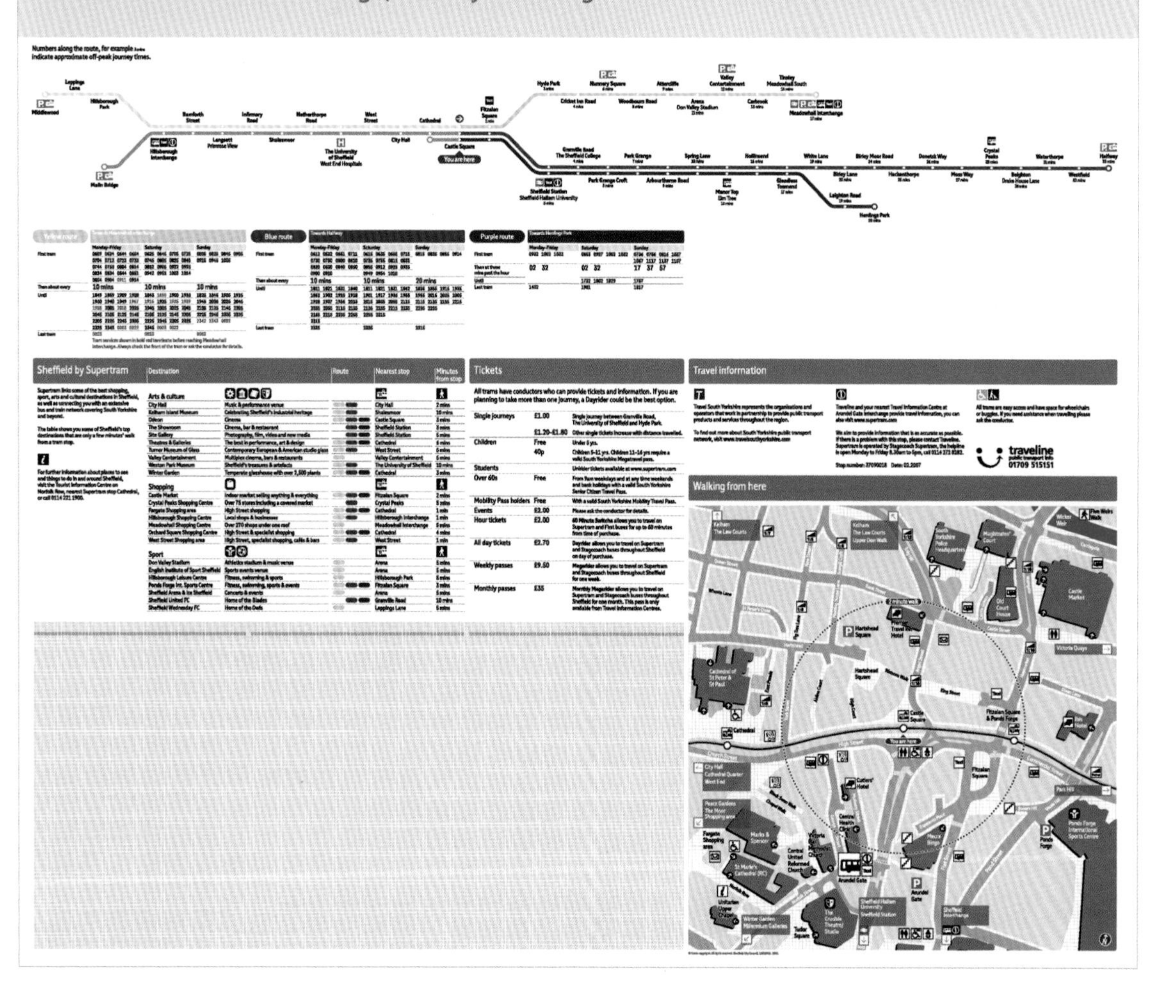
Blue route
Purple route
Sheffield by Supertram
Tickets
Travel information
Walking from here
traveline
01709 515151

03.1.4
多重项目

“三月设计”2009年（Honnunar-Mars）是雷克雅未克有史以来举行的第一次设计节。

这是为世界著名的设计而举办的为期4天的设计节，其视觉标志系统基于两种关键颜色：洋红色与蓝绿色——两种颜色叠印创造出紫色。故意让两种颜色错边，将其产生的效果应用于节日标题及大型色彩块上。时刻表包含在折叠式插页节目单内，使用彩条来表示各种项目。

时刻表将所有文字信息移至左右边缘，使四大中心板块变得与各种颜色条一样抽象，表现了项目的密集程度。这样就可以很容易地看出每一项目的持续时间，以及项目是在上午、下午、晚上或是全天进行。

▸ 01

设计者：Atelier Atli Hilmarsson
作品名：*HonnunarMars (designMarch 2009)*
节日标志与时间表设计

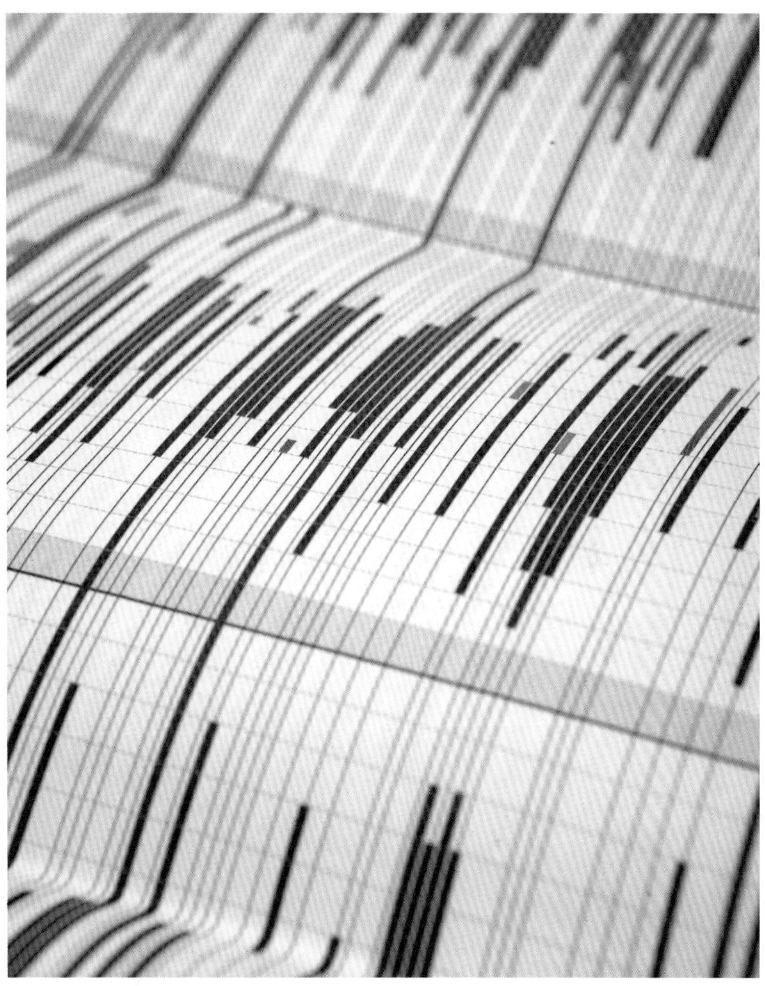

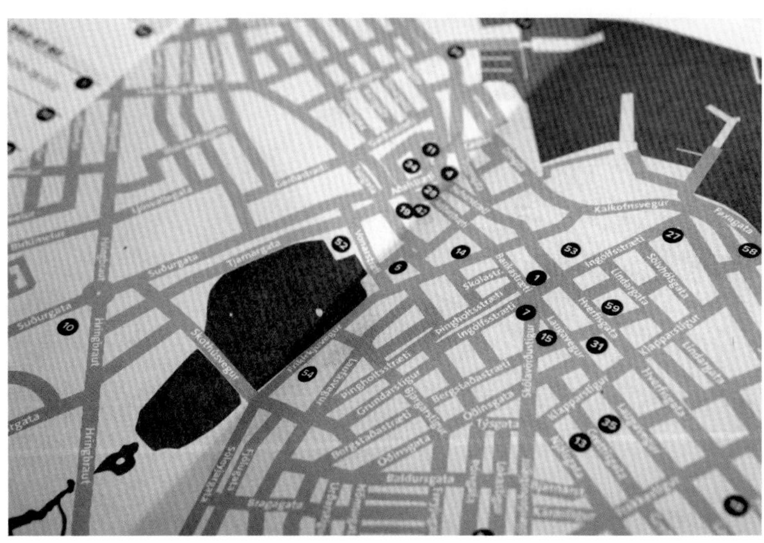
Kalkofnsvegur
Ingólfsstræti
Sölvhólsgata
Lindargata
Hverfisgata
Klapparstígur
Laugavegur
Skólavörðustígur
Bankastræti
Skólastr.
Þingholtsstræti
Grundarstígur
Bergstaðastræti
Óðinsgata
Týsgata
Baldursgata
Skothúsvegur
Suðurgata
Tjarnargata
Hringbraut
1
7
10
13
14
15
27
31
53
54
58
59

03.1.5
线型项目

为赞美荷兰设计传奇人物维姆·克劳威尔（Wim Crouwel），在伦敦设计博物馆举行了大型展览，参与此次活动的英国领先的设计工作室创作了一套限量版海报。“凯特利吉·列文”设计工作室制作的海报采用了艺术及文化活动列表，展示了维姆·克劳威尔与市立博物馆及凡艾伯美术馆共事的大部分内容。海报列出整段展览时间内维姆·克劳威尔将在伦敦展出的主要文化展，显示出展览的内容，并将96天内的展览项目制作成日历。

▸ 01

设计者：Cartlidge Levene
作品名：*A Graphic Odyssey Poster*
摄影：Marcus Ginns
限量版丝印海报

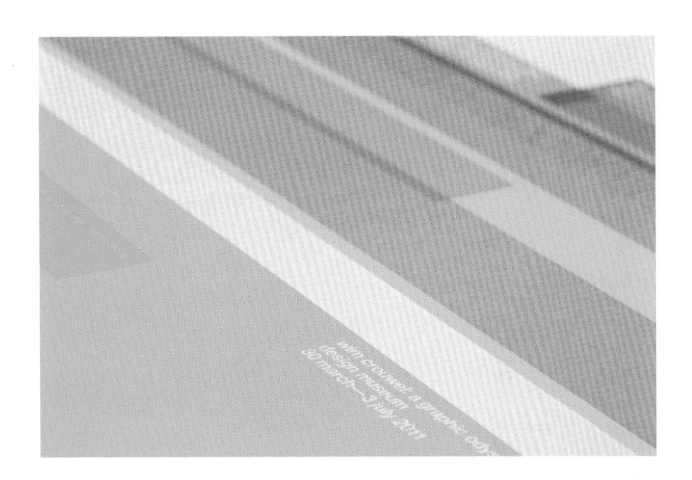
wim crouwel: a graphic
design museum
30 march—3 july 2011

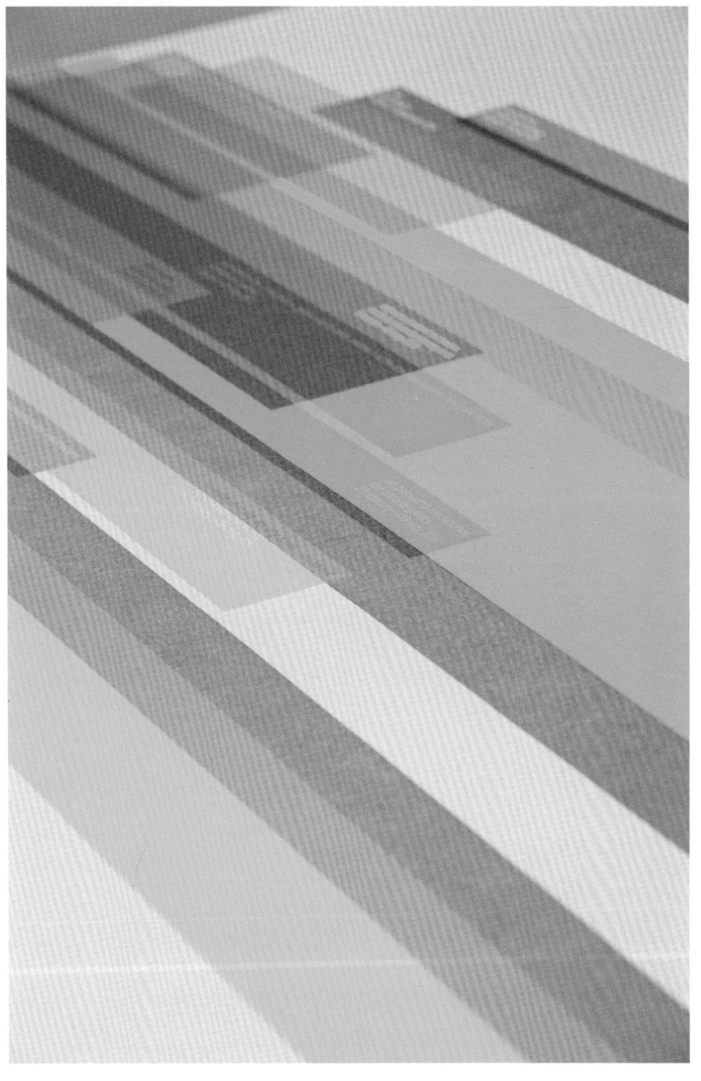

03.1.6
简洁性与复杂性

挂在“Sliver”美术馆中的有关2009年春季讲座的海报保持了信息的清晰与简洁，它简单地采用红色印制于黄色板块上，所有日期均与月份对齐，演讲者的名字则叠放在月份下方。

《7月24日的不在场证明》是纽约《时代》杂志上一篇有关美国艺术教授哈桑・伊拉希（Hasan Eiahi）一项目可视化设计的文章。图像以其个人网页数据信息为基础，呈现了他在2007年的行踪。他个人信息的公布旨在回应他被错误地列在FBI的恐怖分子监视名单中。日期提供了伊拉希在某一特定时刻的所在位置信息，并可以看到他在哪些日子具有无懈可击的不在场证明。

01

▸ 01

设计者：Paulus M. Dreibholz
作品名：*Sliver Lecture Series 2009*
活动海报

▸ 02

设计者：Von B und C
作品名：*24/7 The Alibi 2007*
纽约《时代》杂志专栏

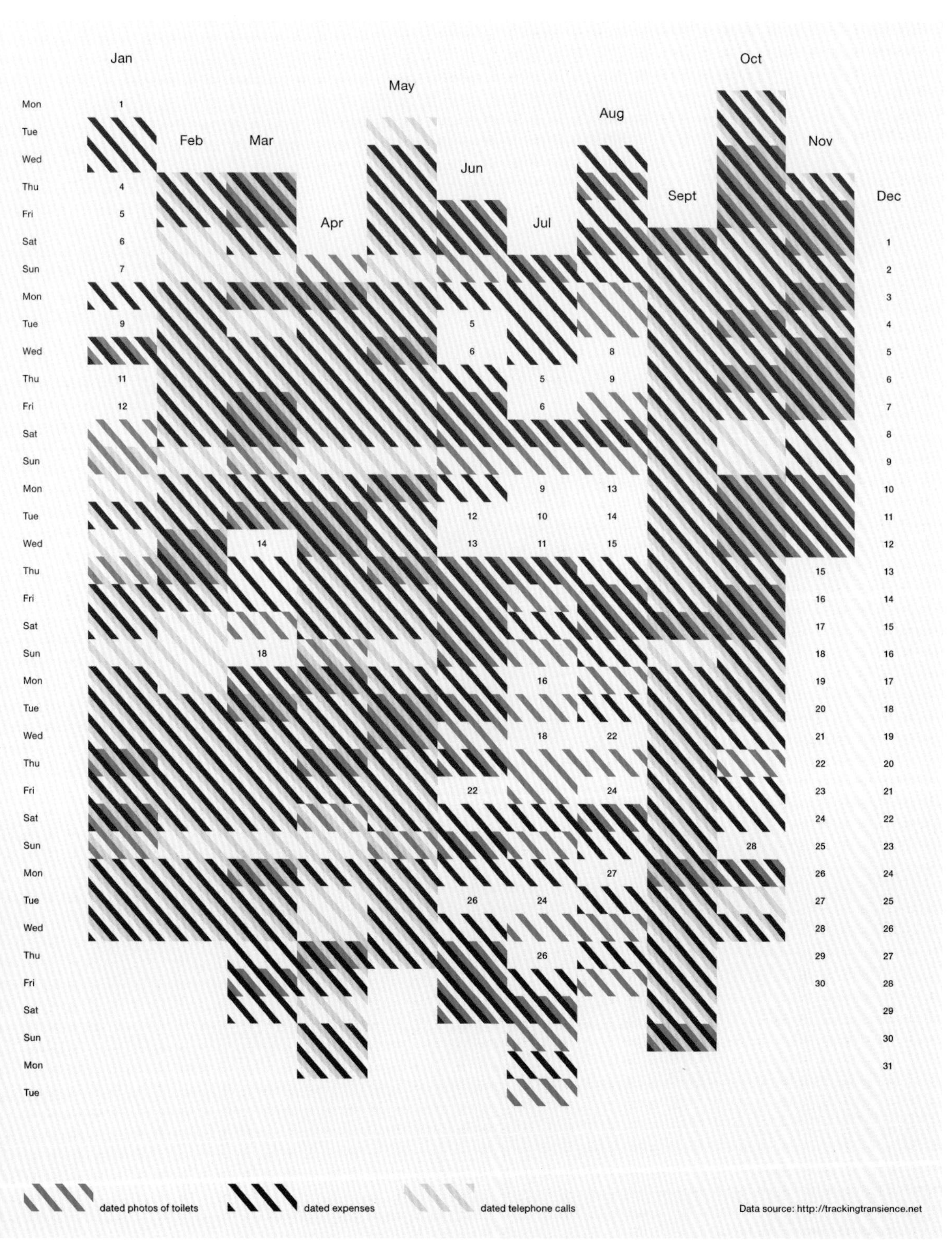
Jan
Feb
Mar
Apr
May
Jun
Jul
Aug
Sept
Oct
Nov
Dec
Mon
Tue
Wed
Thu
Fri
Sat
Sun
dated photos of toilets
dated expenses
dated telephone calls
Data source: http://trackingtransience.net

02

03.2
列表

03.2.1
连续时间

由“Ich & Kar”公司设计的《一条》是一款粘贴式壁纸，可以帮助小朋友学习乘法表。这没有什么古怪或陌生的——仅仅是清晰和没有分散的信息。

“Stapelberg &Fritz”设计公司为“斯图加特设计中心”制作的系列活动传单以大型的裁剪数字为特色，数字显示的是每一传单所表明的月份。通过裁剪第一个与最后一个数字来暗示不间断的活动，而传单所示仅为整个活动的一部分。虽然每一设计都相似，但对每一传单上的数字均进行了不同的调整和裁剪。

01

▸ 01

设计者：Ich & Kar
作品名：*One Strip*
粘贴式壁纸

▸ 02

设计者：Stapelberg & Fritz
作品名：*Design Center Stuttgart*
系列活动宣传单

Design Center Stuttgart

Termine
September
bis
Dezember 2010

Design Center Stuttgart
Regierungspräsidium Stuttgart
Haus der Wirtschaft
Willi-Bleicher-Straße 19
70174 Stuttgart

T +49(0)711-123 2781
design@rps.bwl.de
www.design-center.de

Design Center
Stuttgart

Termine
Januar
bis
März 2011

Vortragsreihe
Design Innovationsmotor
für den Mittelstand

IHK Südlicher Oberrhein,
Freiburg: 26. Januar
IHK Hochrhein Bodensee,
Konstanz: 27. Januar

jeweils von 18.00 bis 15.00 Uhr

Ausstellung »Design zählt«
Mercedes-Benz Design at School
Eröffnung: 17. Februar, 19.00 Uhr
Laufzeit bis 20. März

Ausstellungsreihe Ein( )sichten
Hugo Boss Fashion Award 2011
Vernissage und Preisverleihung
10. März, 19.00 Uhr
Laufzeit: bis 1. April

DesignLese
MANUskript,
Prof. Axel Müller-Schöll
15. März, 19.00 Uhr

Kongress
HMID
Human Machine Interaction Design
18. März, ganztägig

Wettbewerbe

Focus Open
Internationaler Designpreis
Baden-Württemberg 2011
Anmeldeschluss: 25. März

Mia Seeger Preis 2011
Einsendeschluss: 10. Juni

Design1st Beratung
Termininfos online
www.design-center.de

Ort für Ausstellungen,
Veranstaltungen und Beratung
Stuttgart, Haus der Wirtschaft
Baden-Württemberg
Öffnungszeiten täglich
außer sonn- und feiertags
von 11.00 bis 18.00 Uhr

Vormerken:
Wettbewerbe und Ausstellung

Focus Open
Internationaler Designpreis
Baden-Württemberg
und Mia Seeger Preis 2011
Ludwigsburg,
Werkzentrum Weststadt
Preisverleihung am 14. Oktober
Laufzeit der Ausstellung
bis 27. November

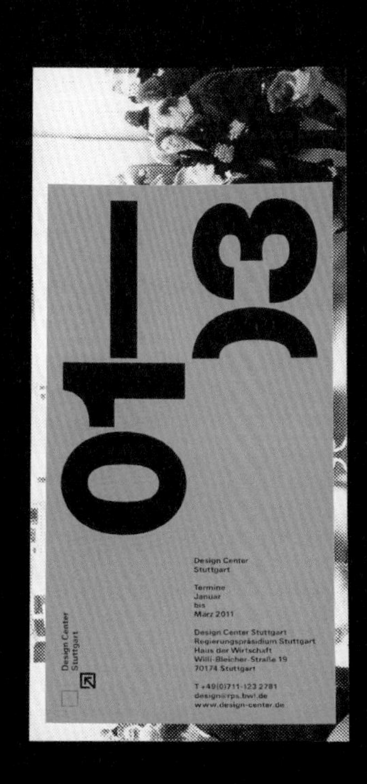

02

03.2.2
日期与时间

为丹麦男装店“Minimum”设计的一系列A5传单及宣传海报，每一幅都是纯粹的字体排版。《1 440》宣传了一项商场活动，活动期间，男装店将24小时营业（1 440分钟）。《1.minimum》是为庆祝该商店开业一周年而设计。

采用有活力的橙色作为商店的促销颜色。丹麦大多数商店促销时都采用黄色，而在英国则采用的是红色。而“Minimum”则混合使用这两种颜色。1998年的夏季促销海报在字母“S”与数字“8”和“9”的相似性上大做文章。

01

02

▸ 01

设计者：Struktur Design
作品名：*1,440 minimuminutter*
24小时购物活动海报

▸ 02

设计者：Struktur Design
作品名：*1.mininum*
商店一周年纪念海报

▸ 03

设计者：Struktur Design
作品名：*Pre-view*
宣布两场时装预展

▸ 04

设计者：Struktur Design
作品名：*1998 summer sale*
1998夏季促销海报

minimum

pre-
view 30-31.12.98 –
02.01.99

03

1
9
9ummer
8ale

du indbydes
hermed til for-
præmiere på
summer sale
1998

torsdag 25. juni
10.00 – 19.00

fredag 26. juni
10.00 – 19.00

lørdag 27. juni
9.30 – 16.00

minimum

04

03.2.3
垂向层序

由威利・孔茨（Willi Kunz）设计的两幅海报的垂直格式通过信息结构来加以强调。哥伦比亚建筑保护规划采用长而窄的栏来呈现根本的梯度系统。事件日期是海报最显著的特征，是其结构的支柱，旁边仅提及将要进行活动的月份。相比之下，哥伦比亚大学建筑与规划研究生院则充分采用白色空间，而粗体黄色日期依旧是海报的焦点。

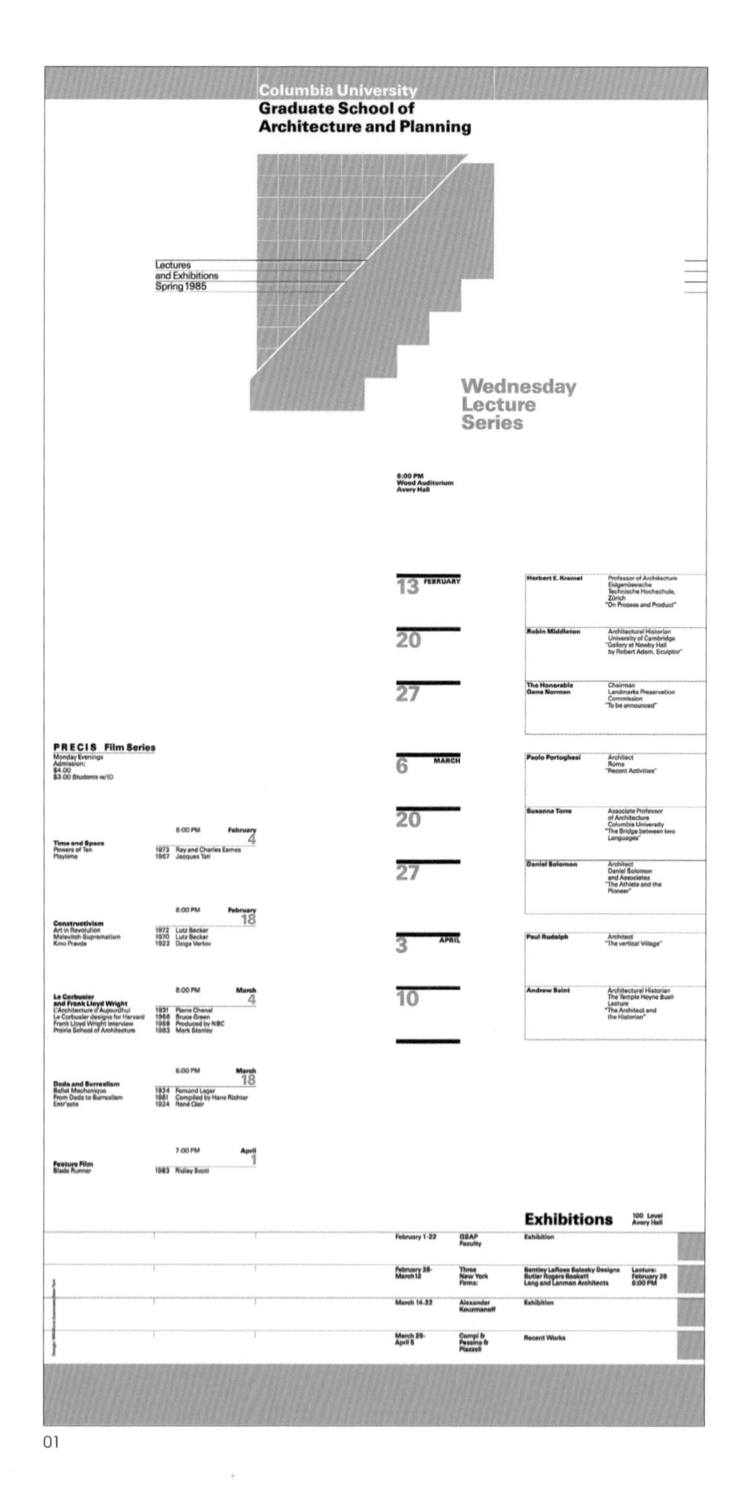

01

▸ 01

设计者：Wilti Kunz
作品名：*Columbia University Gradcate School of Arachitecture and Planning*
海报

▸ 02

设计者：Wilti Kunz
作品名：*Columbia Architecture Planning Preservation*
海报

Columbia Architecture Planning Preservation

February
01 02 WTC Forum

Introduction
Bernard Tschumi, Dean
Mary McLeod
Laurie Hawkinson

The World Trade Center and the phenomenon of 'tallest' towers
Bill Faschan
Richard Muller
Carol Willis
Paul Byard (moderator)

Real estate, development, finance
Charles Bagli
Richard Plunz
David Stark
Carl Weisbrod
Elliott Sclar (moderator)

Beyond Finance: Infrastructure, Ecology, and Everyday Life
Liz Abzug
Cynthia Rosenzweig
Andrew Ross
Marilyn Taylor
Sharon Zukin
Stan Allen/Gwen Wright (moderators)

Global implications Trauma and memory
Mindy Fullilove
Ray Gastil
Andreas Huyssen
Kevin Kennon
Mark Wigley (moderator)

Image and Spectacle
Marshall Berman
Robert Stern
Bernard Tschumi
Joan Ockman (moderator)

Roundtable discussion
Mary McLeod (moderator)

Lectures

6:30pm
Wood Auditorium
Avery Hall

Doors open to the general public 6:15pm

February
28 Ando
Thursday, February 28
1:00pm

Tadao Ando
Architect, Japan

Recent work

28 Koolhaas
Thursday, February 28
4:00pm

Rem Koolhaas
OMA, Rotterdam and London

The Lagos Project
Response:
Manthia Diawra, NYU

The Sawyer Seminar, directed by Andreas Huyssen, Columbia University

March
13 Mori
Wednesday, March 13

Toshiko Mori
Architect, New York
Robert Hubbard Professor in Practice of Architecture, Harvard GSD

Material/Immaterial

06 Hadid
Wednesday, March 6
6:30pm

Zaha Hadid
Architect, London

Recent work

April
03 Meyers
Wednesday, April 3

Victoria Meyers
Partner,
Hanrahan+Meyers Architects
Adjunct Assistant Professor of Architecture,
Columbia University

four states of architecture

02 Vidler
Tuesday, April 2
6:30pm

Anthony Vidler
Historian/Theorist

Modernism and Autonomy

10 Mendes da Rocha
Wednesday, April 10

Paolo Mendes da Rocha
Architect, São Paulo

Recent work

12 Prouvé
Friday, April 12
4:00–6:30pm

A symposium on the work of Jean Prouvé in conjunction with the exhibition on view in the Arthur Ross Architecture Gallery

17 Barber
Wednesday, April 17

Benjamin Barber
Professor of Civil Society, University of Maryland;
Director, New York office of The Democracy Collaborative

Cosmopolitanism vs. Fundamentalism: The City as Democracy's Forge

Buell Evening Lecture co-sponsored by Skidmore, Owings & Merrill

19 Ito
Friday, April 19

Toyo Ito
Architect, Japan

Recent work

Exhibitions

Spring 2002

January 28–March 14
Lucid Stillness: Photography by Candida Hofer
400 Avery

January 29–February 15
Johannesburg: The Country in the City Photographs by Jodi Bieber
100 Avery Hall

February 1–February 15
WTC Forum
100 Avery Hall

February 18–March 15
Havana: The Photography of Hans Engels
100 Avery

March 25–May 10
Satellite of Love: Vanishing Beauty of Japanese Love Hotels
Curated by Kyoichi Tsuzuki
200 Avery

March 27–May 3
Soho: Photography by Bevan Davies
400 Avery

March 28–May 10
Industrial Alchemy: Radical Pragmatism in the Work of Jean Prouvé
Curated by Evan Douglis and Robert Rubin
Arthur Ross Gallery, Buell Hall

April 1–May 3
Reclaiming the Western American Landscape
Curated by Alan Berger
100 Avery

May 18–June 1
End of Year Student Exhibition
Avery and Buell Hall Galleries

Symposia

Saturday, February 23
10:00am–4:00pm

555 Lerner Hall,
Columbia University

James Marston Fitch Colloquium

Target Architecture: The Role of old Buildings in the Management of Global Conflict

sponsored by the Historic Preservation Program, Columbia University Graduate School of Architecture Planning and Preservation

Friday, April 5
Saturday, April 6

Urban Design: Practices, Pedagogies, Premises

Friday, April 5
6:00–9:30pm
Lighthouse International
111 East 59th Street

Shaping Civic and Public Realms: What is the Role of Urban Design?

Saturday, April 6
9:00am–6:00pm
Wood Auditorium,
Avery Hall,
Columbia University

Urban Design Practices, Urban Design Pedagogies, Urban Design Premises

Moderated public discussion will follow each panel

Sponsored by the Urban Design Program Columbia University Graduate School of Architecture Planning and Preservation; Urban Design Program, Department of Urban Design and Planning Theory, GSD, Harvard University; Van Alen Institute

Design: Willi Kunz Associates, New York

02

03.2.4
日期与数据

《设计：彻底的》是为在利兹都市大学建筑/景观与设计学院举行的演讲讨论会而设计的海报。由于海报将张贴在学校的通告栏中，因而视觉上它必须在大批宣传海报和公告中引人注目。使用粗体、有趣的排版来表达讲座的性质。

在一系列持续印刷的印刷品中，“Mark Bloom of Mash Creative”首次为《对我而言这才是真正有用的》系列海报设计了4份主题海报，延伸其品牌系列“State of the Obvious”背后的哲学与思考。海报包括：《格林威治标准时间》《表情符号》《印制尺寸》及《公制转换》。

01

▸ 01

设计者：Bibliotheque
作品名：*Design:Inside Out*
讲座论坛海报

▸ 02

设计者：Mash Creative
作品名：*This is my Really Useful*
《表情符号》主题海报

▸ 03

设计者：Mash Creative
作品名：*This is my Really useful*
《表情符号》主题海报

▸ 04

设计者：Mash Creative
作品名：*This is my Really useful*
《格林威治标准时间》主题海报

▸ 05

设计者：Mash Creative
作品名：*This is my Really useful*
《印制尺寸》主题海报

02

03

This is my
y Really
Useful
ch Greenwich
e Mean Time
Poster.™
Addis Ababa + 3hrs
Adelaide* + 10.30hrs
Aden + 3hrs
Algiers + 1hr
Almaty + 6hrs
Amman + 2hrs
Amsterdam + 1hr
Anadyr - 1hr
Anchorage - 9hrs
Anarka + 2hrs
Antananarivo + 3hrs
Asuncion* - 3hrs
Athens + 2hrs
Atlanta - 5hrs
Auckland* + 13hrs
Badgad + 3hrs
Bangkok + 7hrs
Barcelona + 1hr
Beijing + 8hrs
Beirut + 2hrs
Belgrade + 1hr
Berlin + 1hr
Bogota - 5hrs
Boston - 5hrs
Brasilia* - 2hrs
Brisbane + 10hrs
Brussels + 1hr
Bucharest + 2hrs
Budapest + 1hr
Buenos Aires - 3hrs
Cairo + 2hrs
Canberra* + 11hrs
Cape Town + 2hrs
Caracas - 4.30hrs
Casablanca +/- 0hrs
Chatham Islands* + 13.45hrs
Chicago - 6hrs
Columbus - 5hrs
Copenhagen + 1hr
Darwin + 9.30hrs
Denver - 7hrs
Detroit - 5hrs
Dhaka + 6hrs
Dubai + 4hrs
Dublin +/- 0hrs
Edmonton - 7hrs
Frankfurt + 1hr
Geneva + 1hr
Guatemala - 6hrs
Halifax - 4hrs
Hanoi + 7hrs
Harare + 2hrs
Havana - 5hrs
Helsinki + 2hrs
Hong Kong + 8hrs
Honolulu - 10hrs
Houston - 6hrs
Indianapolis - 5hrs
Islamabad + 5hrs
Istanbul + 2hrs
Jakarta + 7hrs
Jerusalem + 2hrs
Johannesburg + 2hrs
Kabul + 4.30hrs
Kamchatka + 11hrs
Karachi + 5hrs
Katmandu + 5.45hrs
Khartoum + 3hrs
Kingston - 5hrs
Kiritimati + 14hrs
Kolkata + 5.30hrs
Kuala Lumpur + 8hrs
Kuwait City + 3hrs
Kyiv + 2hrs
La Paz - 4hrs
Lagos + 1hr
Lahore + 5hrs
Lima - 5hrs
Lisbon +/- 0hrs
London +/- 0hrs
Los Angeles - 8hrs
Madrid + 1hr
Managua - 6hrs
Manila + 8hrs
Melbourne + 11hrs
Mexico City - 6hrs
Miami - 5hrs
Minneapolis - 6hrs
Minsk + 2hrs
Montevideo* - 2hrs
Montgomery - 6hrs
Montreal - 5hrs
Moscow + 3hrs
Mumbai + 5.30hrs
Nairobi + 3hrs
Nassau - 5hrs
New Delhi + 5.30hrs
New Orleans - 6hrs
New York - 5hrs
Oslo + 1hr
Ottawa - 5hrs
Paris + 1hr
Perth + 8hrs
Philadelphia - 5hrs
Phoenix - 7hrs
Prague + 1hr
Reykjavik +/- 0hrs
Rio de Janeiro* - 2hrs
Riyadh + 3hrs
Rome + 1hr
San Francisco - 8hrs
San Juan - 4hrs
San Salvador - 6hrs
Santiago* - 3hrs
Santo Domingo - 4hrs
São Paulo* - 2hrs
Seattle - 8hrs
Seoul + 9hrs
Shanghai + 8hrs
Singapore + 8hrs
Sofia + 2hrs
St. Johns - 3.30hrs
Stockholm + 1hr
Suva* + 13hrs
Sydney* + 11hrs
Taipei + 8hrs
Tallinn + 2hrs
Tashkent + 5hrs
Tegucigalpa - 6hrs
Tehran + 3.30hrs
Tokyo + 9hrs
Toronto - 5hrs
Vancouver - 8hrs
Vienna + 1hr
Vladivostok + 10hrs
Warsaw + 1hr
Washington DC - 5hrs
Winnipeg - 6hrs
Yangon + 6.30hrs
Zagreb + 1hr
Zürich + 1hr
PRINT-PROCESS

04

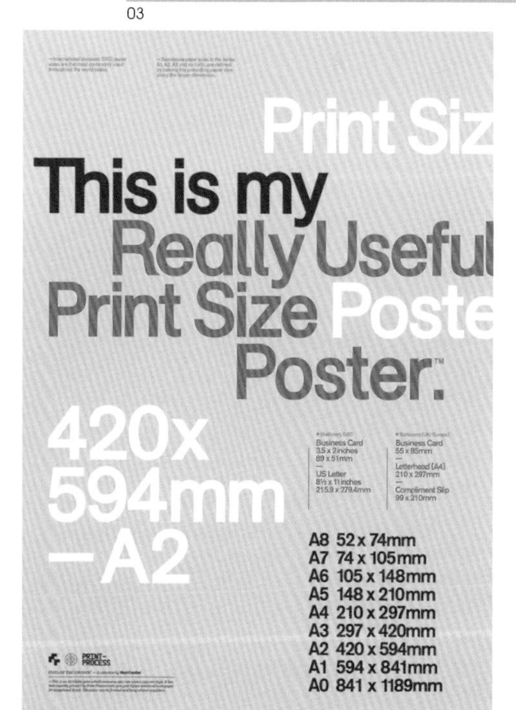

05

03.2.5
线与边框

《创意2007》是中东创意发布会用A5纸印制的简介小册子。为期4天的活动计划包括研讨会与会议谈判，具有层次结构的支配权重有助于对其进行区别。

伦敦莫莫“Kemia”酒吧的海报简单明了地列出所有现场音乐活动月份及日期。所有信息都被印制成3种语言。

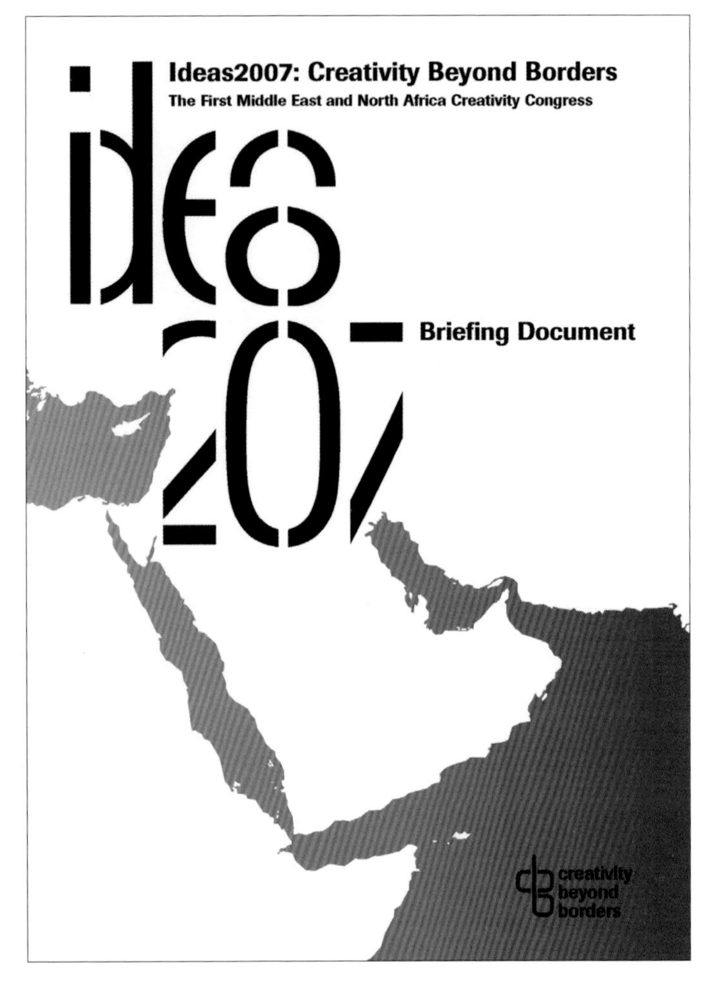

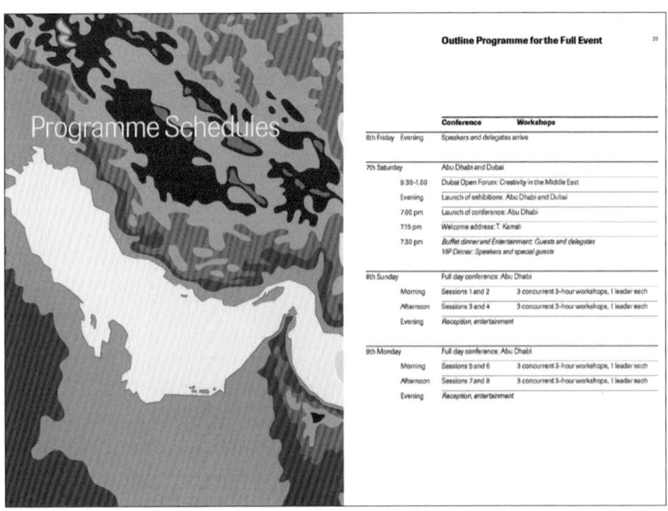

01

▸ 01

设计者：Struktur Design
作品名：*Ideas 2007*
简报与活动安排

▸ 02

设计者：Ich &Kar
作品名：*Kemia Bar*
活动海报

january 2002
janfie 2002 at kemia bar
janvier 2002

food every evening from 7 PM till late
valet parking, if you need any further information, please don't hesitate to contact us
0207 434 20 11 / 25 HEDDON STREET - W1 LONDON / momoresto@aol.com

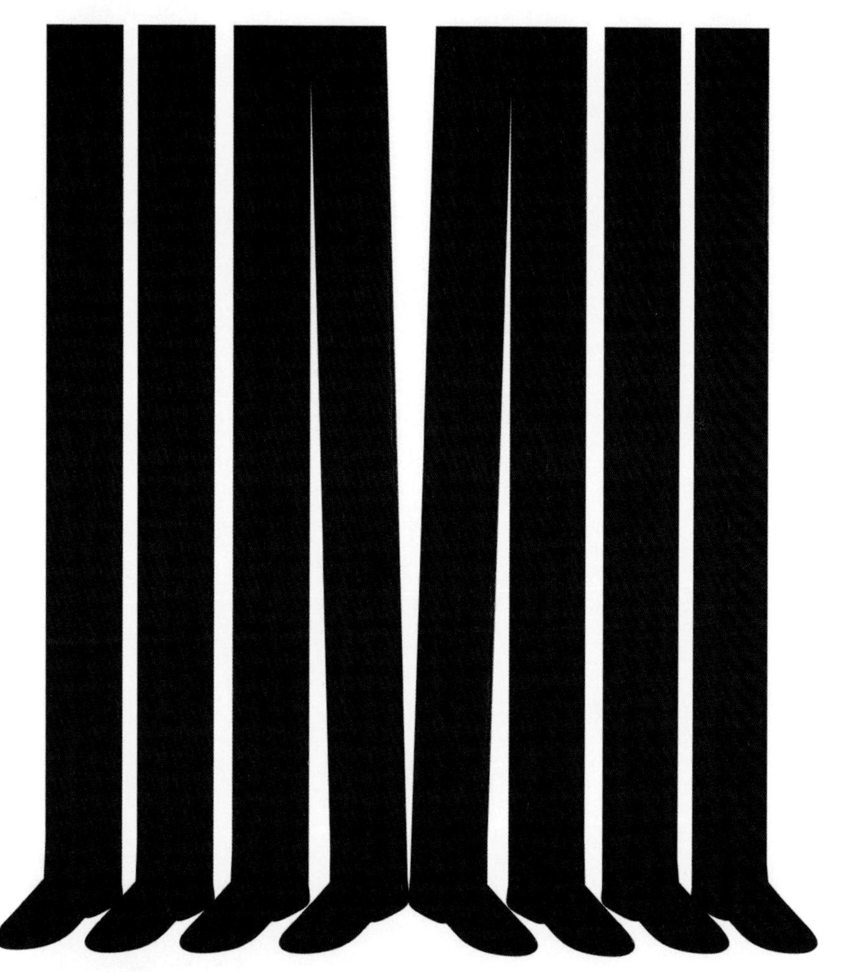

monday, tuesday, wednesday- 21:00
lethnin, thlata, larabaa- 21:00
lundi, mardi, mercredi- 21:00

live music

thursday, friday, saturday
l'khemiss, djemaa, sebte
jeudi, vendredi, samedi

momo's music

monday / lethnin / lundi	tuesday / thlata / mardi	wednesday / larabaa / mercredi	thursday / l'khemiss / jeudi	friday / djemaa / vendredi	saturday / sebte / samedi
	01 sorry closed	**02** open	**03** resident dj boris	**04** resident dj boris	**05** resident dj boris
07 roberto pla and his guests salsa dj simon laurence	**08** kanda bongo man congolese soukouss dj boris	**09** ray gaskins jazz funk guest dj	**10** resident dj boris	**11** resident dj boris	**12** resident dj boris
14 roberto pla and his guests salsa dj simon laurence	**15** chartwell dutiro mbira from zimbabwe dj boris	**16** mandisa soul guest dj	**17** resident dj boris	**18** resident dj boris	**19** resident dj boris
21 roberto pla and his guests salsa dj simon laurence	**22** karuna & sanjeev urdu and persian poetry dj boris	**23** don blackman funk soul guest dj	**24** resident dj boris	**25** resident dj boris	**26** resident dj boris
28 roberto pla and his guests salsa dj simon laurence	**29** another fine day world fusion dj boris	**30** lore trip-hop guest dj	**31** resident dj boris		

02

03.2.6 超大号日期

“La Casa Encendida”是马德里的社交、资源及文化中心，这里经常举办各种各样的现代艺术展及教育活动。

其小册子的组成元素为1种基色和5种混合色，5种颜色分别代表该机构的5个部门。每月的封面都印有一个巨大而有活力的绿色日期来充当主导。

01
设计者：Base Design
作品名：*La Casa Encendida*
每月宣传封面

03
10
LA CASA ENCENDIDA

04
10
LA CASA ENCENDIDA

05
LA CASA ENCENDIDA

06
LA CASA ENCENDIDA

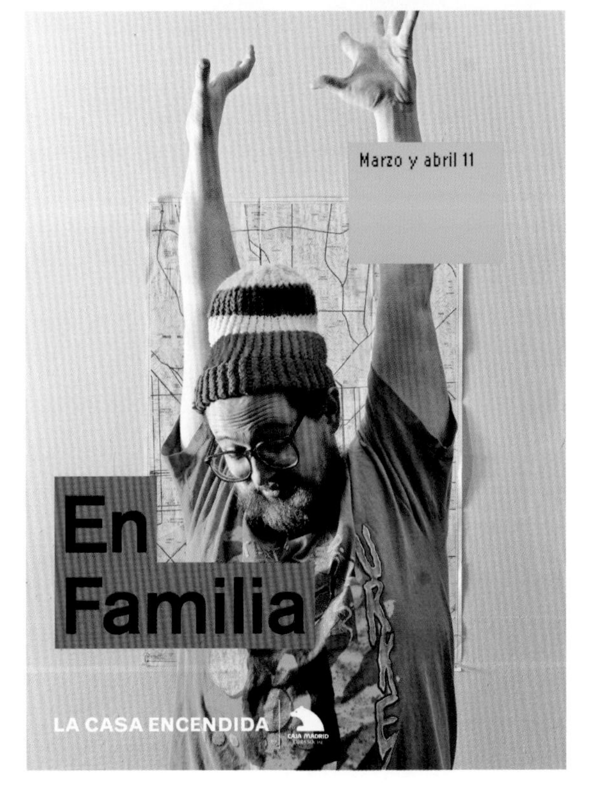
Marzo y abril 11
En
Familia
LA CASA ENCENDIDA

Verano
en La Casa
Julio y agosto 10
LA CASA ENCENDIDA

03.2.7
分片

《Distribute! 2》是为在德国斯图加特符腾堡美术馆举行的一项影视艺术活动设计的风琴折式传单，对粗体日期及标题加以分割。风琴式折叠是相互交错的，当传单合起时每一面的某部分仍然是可见的。活动的标题填满折叠单的一侧，并在背面进行描述。当传单展开时标题呈分片状并显示出日期，一起显示的还有活动期间具体事项的详细信息。

▸ 01

设计者：Stapelberg & Fritz
作品名：*Distribute! 2*
风琴折式传单

5./
6.
2
NOV
10
DISTRIBUTE! 2
2010

VIDEOKUNST 5./6.
DISTRIBUTE! 2

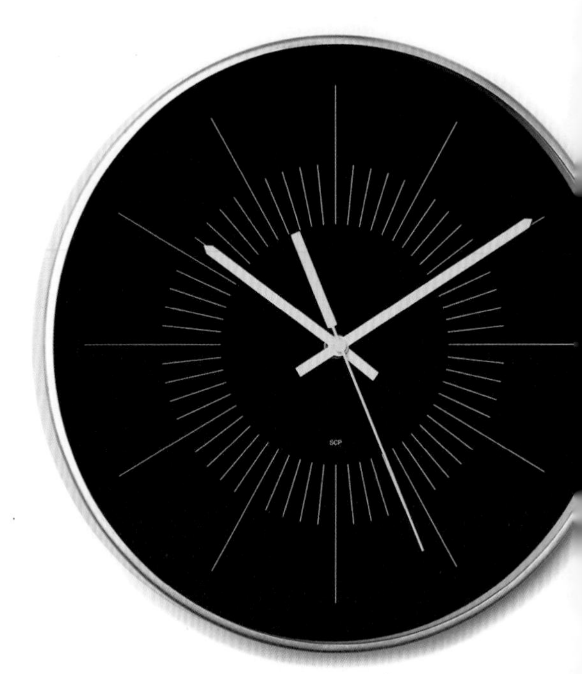

04.
年表

04.1 日历

04.1.1 胶片上的日期 118
04.1.2 枪靶与积木 122
04.1.3 周视图 124
04.1.4 成百上千 126
04.1.5 规则与细节 128
04.1.6 数字序列 130
04.1.7 切割成片 132
04.1.8 多项选择 134
04.1.9 挂历策划 136

04.2 日记簿

04.2.1 年视图 138
04.2.2 版式轮廓 140
04.2.3 彩色序列 142
04.2.4 可用性 144

04.3 钟表

04.3.1 模拟数字 146
04.3.2 数码数字 148
04.3.3 原子钟 150
04.3.4 时钟应用程序 152
04.3.5 探索时间 156
04.3.6 组织时间 158

04.1
日历

04.1.1
胶片上的日期

为纪念1997年英国交还中国香港主权，设计者手工制作了16毫米长的胶片，上面印有凸版数字。每天24帧，用大小和深浅上发生增减的数字来对每小时进行描述，以获得“上升”和“沉没”的效果。这些数字在西边升起，东边落下，直至7月1日政权完成交接的这一时间点，此后数字从东边升起，西边落下。

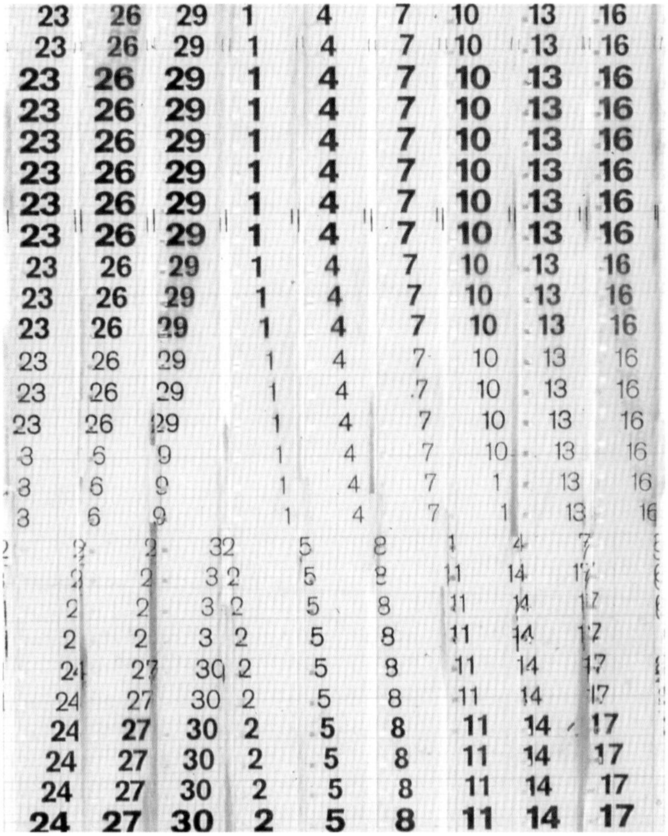

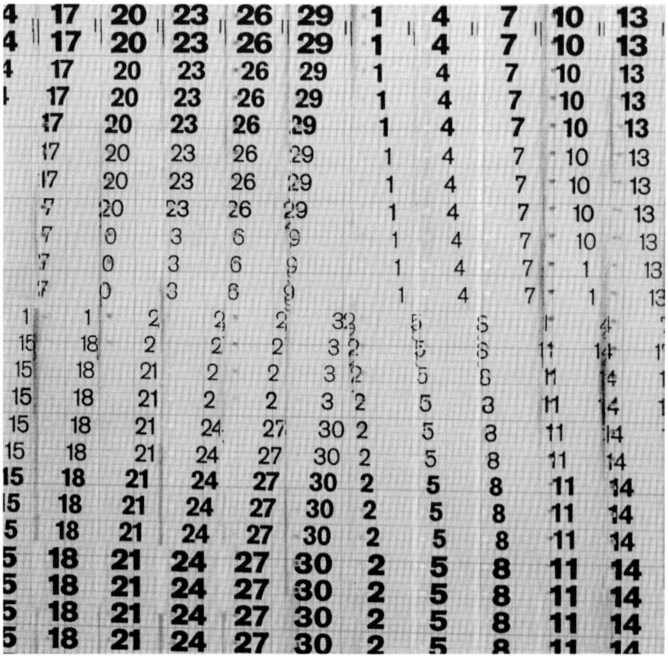

01

设计者：Acme Studios
作品名：*Hong Kong Handover 1997*
16毫米凸版印刷
胶片带

01

04.1.1
胶片上的日期

为纪念1997年英国交还中国香港主权，设计者手工制作了一个纸灯笼，结合丝印、橡皮图章及木版印刷在内的数种印刷法，重叠印制于透明宣纸的两面。灯笼上绘制有1997年的阳历和阴历，自西方新年至1998年二月的春节。每一周为一个单独的七边形，用火柴手工筑成其内部结构。随后七边形被逐个堆叠起来，每移位一次便旁置出一天以追随月球的位置。最后七边形被手工缝制到一起形成灯笼，这种灯笼在内部照明。

同样的情况也可以在设计师的胶片带中见到。数字凸印于16毫米胶片上，投影并拍摄视频后在电视显示器上播放。

02

▸ 02

设计者：Acme Studios
作品名：*Hong Kong Handover 1997*
纸灯笼日历

▸ 03

设计者：Acme Studios
作品名：*Hong Kong Handover 1997*
凸版印刷于16毫米胶片，投影并拍摄视频

03

04.1.2
枪靶与积木

《靶心》是为期一年的日历，大致包含2007—2008年10米气动手枪射击练习的情况。每周印制最近一场比赛使用过的目标靶，且每个靶上均有两次射击。所有目标靶都经国际射击联盟（ISSF）认可。每个靶都采用红、黑两种颜色进行叠印，顶部是大的红色周数，周内的每一天穿过中心位置。

意大利高级纸张主要供应商费德欧尼公司的一种便利贴式日历，每天为一页，每月一种颜色，刺穿的折叠数字创造出一种巧妙的三维视感。

01

▸ 01

设计者：Hörður Lárusson
作品名：*Bullseye*
10米气手枪训练日历

▸ 02

设计者：Studio 8 Desgin
作品名：*Fedrigoni*
日历积木

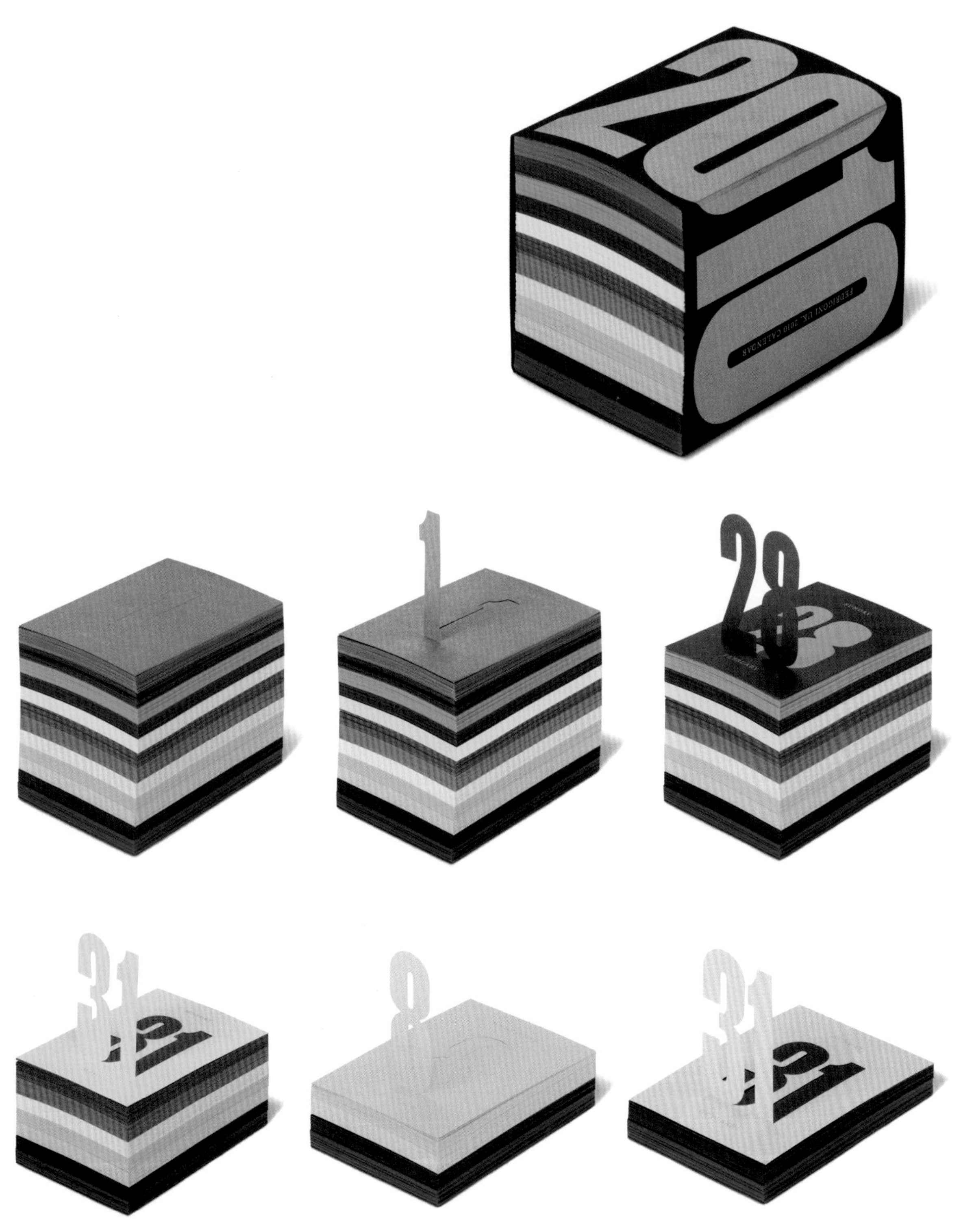

02

04.1.3
周视图

费德欧尼公司的促销台历（分两种语言）是一款按色彩顺序陈列的周视图提升式卡片。设计者为数字自定义了一种字体，银白色的数字印制于各色块上。这种日历创造出一种新颖、富有结构性的方案来处理传统的台历，依靠纸的质量来形成稳固的基底。

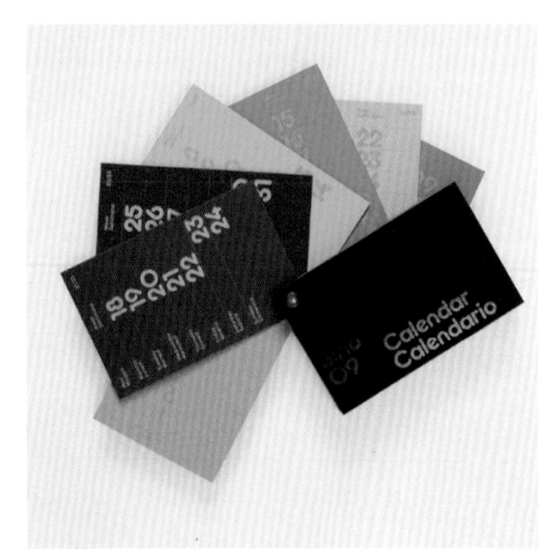

▶ 01
设计者：Design Project
作品名：*Fedrigoni*
台历

29/53
18
19
July
Luglio
Week
Settimana
31/53
Monday
Lunedì
27
Tuesday
Martedì
28
Wednesday
Mercoledì
29
Thursday
Giovedì
30
Friday
Venerdì
31
June
Giugno
Week
Settimana
26/53
Monday
Lunedì
22
Tuesday
Martedì
23
Wednesday
Mercoledì
24
Thursday
Giovedì
25
Friday
Venerdì
26
Saturday
Sabato
27
Sunday
Domenica
28
Calendar
Calendario
Calendar
Calendario

04.1.4
成百上千

《秒》是“马什创意”设计的A1限量版日历三部曲中最早的作品。日历顶部切割成片的标题文本受古老的模拟翻页时钟影响而设计。每一月份以简洁的栏式布局，以略带随意的方式来放置，得以与标题的样式相呼应。

《分》由525 600分钟组成，平均每月分钟数为43 800左右。该设计以年的版式呈现，每一月均被分解成分钟数陈列于每月份名称的下部。日历采用黑暗中会发光的油墨印制，使得该设计在晚上看起来具有与白天完全不同的效果。采用恒星（星）时的理念，突出显示的月份像星星闪烁。

《时》为三部曲中的结束部分，用黑色、金色及透明清漆丝网印刷。

01

▸ 01

设计者：Mash Creative
作品名：*Seconds*
31 556 926 秒

▸ 02
▸ 03

设计者：Mash Creative
作品名：*Minutes*
525 600分

▸ 04

设计者：Mash Creative
作品名：*Hours*
8 784时

Twenty Eleven:
_Five hundred
and twenty
five thousand
_six hundred
minutes.
January
February
March
April
May
June
July
August
September
October
November
December
Produced by MashCreative
www.mashcreative.co.uk

02

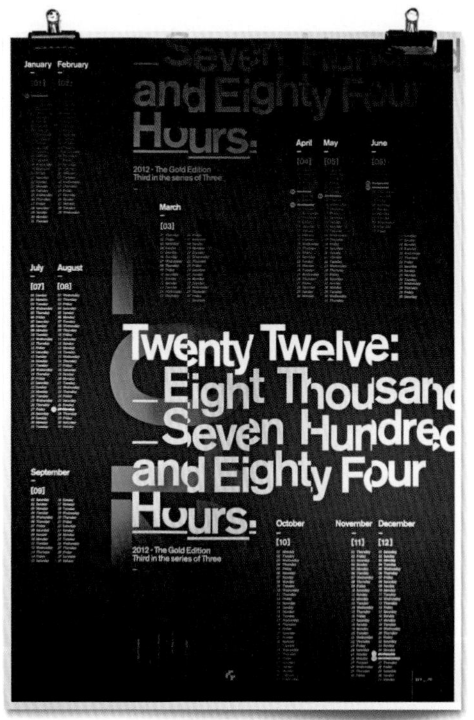
January
February
April
May
June
March
July
August
September
Twenty Twelve:
_Eight Thousand
_Seven Hundred
and Eighty Four
Hours.
2012 - The Gold Edition
Third in the series of Three
October
November
December

04

_Five hundred
and twenty
five thousand
_six hundred
minutes.
May
June
July
March
April
October
November
January
February
December
August
September
Produced by MashCreative
www.mashcreative.co.uk

03

04.1.5
规则与细节

“斯图克尔”公司设计的1998年日历以A2海报的形式印制成金属银/蓝和黑色。1周7天以大号白色字体向下垂直排列。紧随其后有12栏，每一栏代表一个月份。除周末采用白色进行突出设计外，其他文本均印制为黑色。

1997年“斯图克尔”公司设计的首款日历采用柔和的金属质感金色印制于非常薄的圣经纸上，以风琴的形式折叠，使用完全真空密封的包装进行发货。第一个板块包括1996年的最后一个月，日历底部为一个超出纸张的显示1998年开端的小块。

01

▸ 01

设计者：Struktur Design
作品名：*1998 Calendar*
A2海报形式

▸ 02

设计者：Struktur Design
作品名：*1997 Calendar*
可折叠伸缩，圣经纸，尺寸594毫米X210毫米

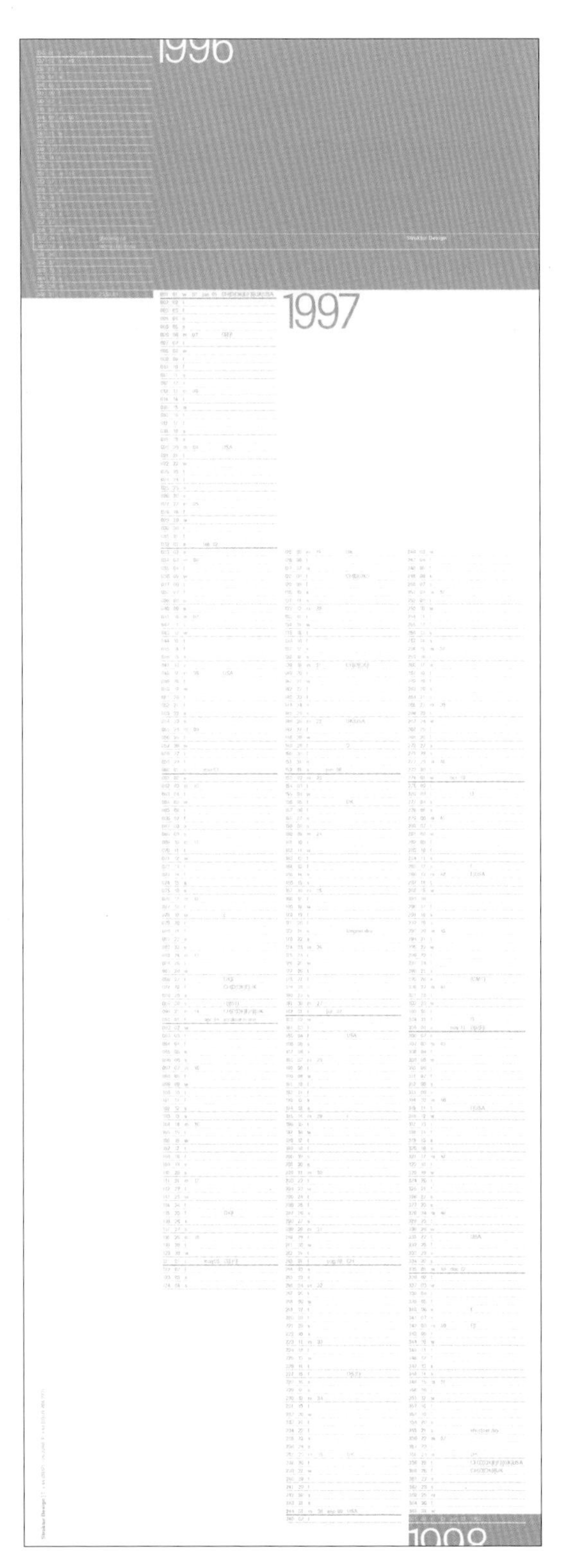

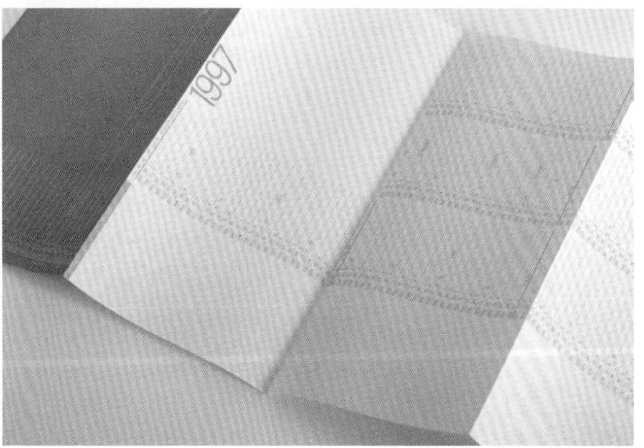

02

04.1.6
数字序列

《分》日志簿的风琴折页面延伸长度约为6米。年度被分解成分钟数，共525 600分钟，采用荧光粉进行印制，并设置每10分钟为一个间隔。分钟数形成的数字长河遍布于日志簿的全长之上，采用一处返回式强调对每月的开始进行标注。在日志簿的每一折痕处添加行间距。小规模数据被套印在黑色的1到365的数字中。周数采用暖灰色加以突出，月份采用浅灰色的较大字体。数字序列的节奏性本质在这一作品的外观上起着关键作用。

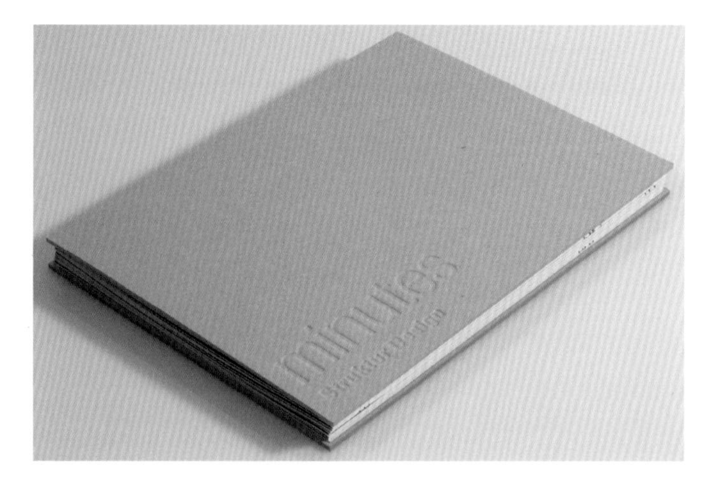

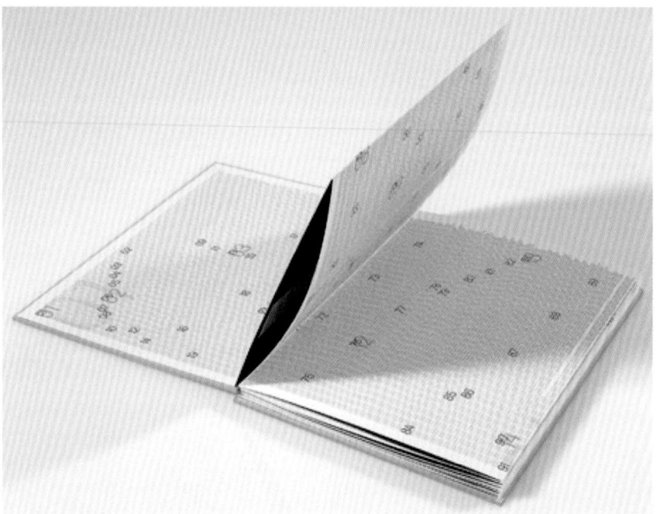

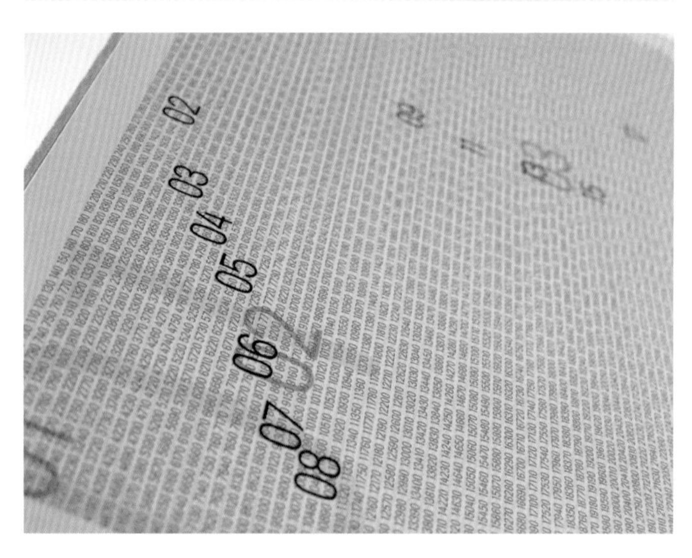

▸ 01

设计者：Struktur Design

作品名：*Minutes*

可伸缩折叠的日记

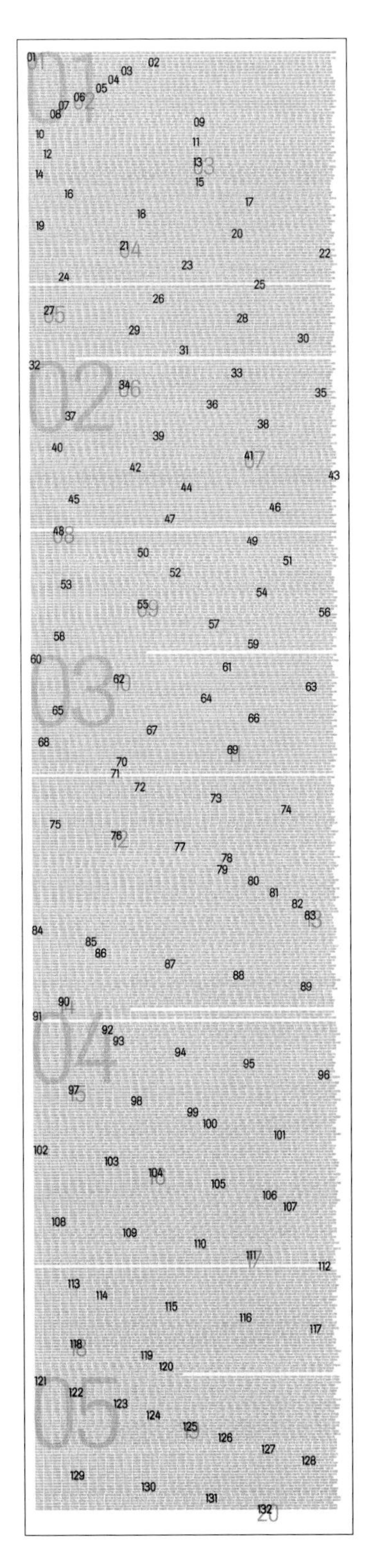

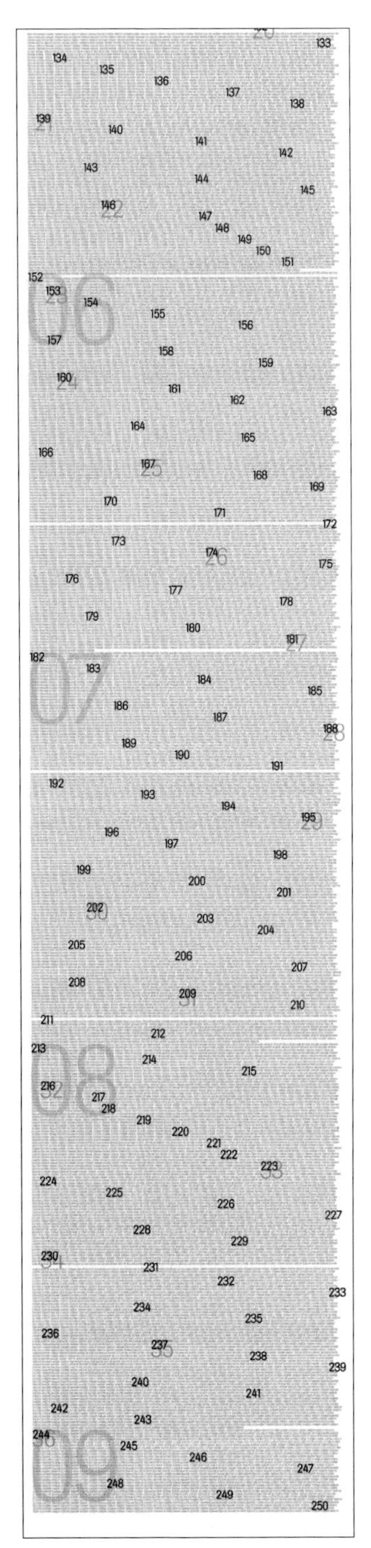

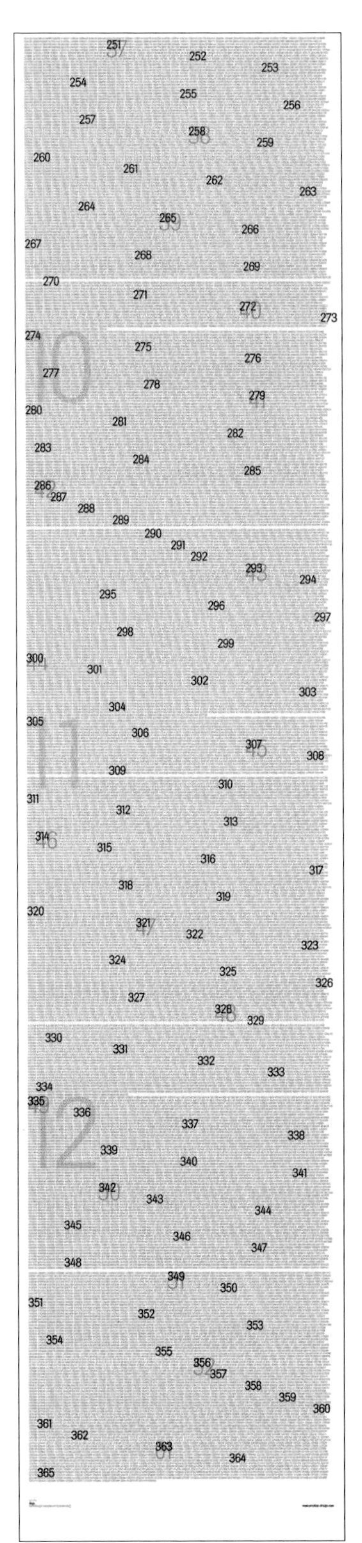

04.1.7
切割成片

设计者仅采用洋红与暖黄两种颜色，结合固墨技术创造出《31日》的各种颜色组合，《31日》是一款31页的壁式日历。通过应用于每页的高光泽UV亮光漆，墨的灵动进一步增强。设计者探究了可读性的极限，每一个数字都被切割成薄条，在某些页面上形成有趣的抽象排版形式。

《52周》被“斯图克尔”公司设计成小型的32页袖珍日记簿，以将01~52周的大号数字切片为特色。每周的两个数字越来越宽且重叠部分越来越多，直至年终数字变成一个抽象的图形。

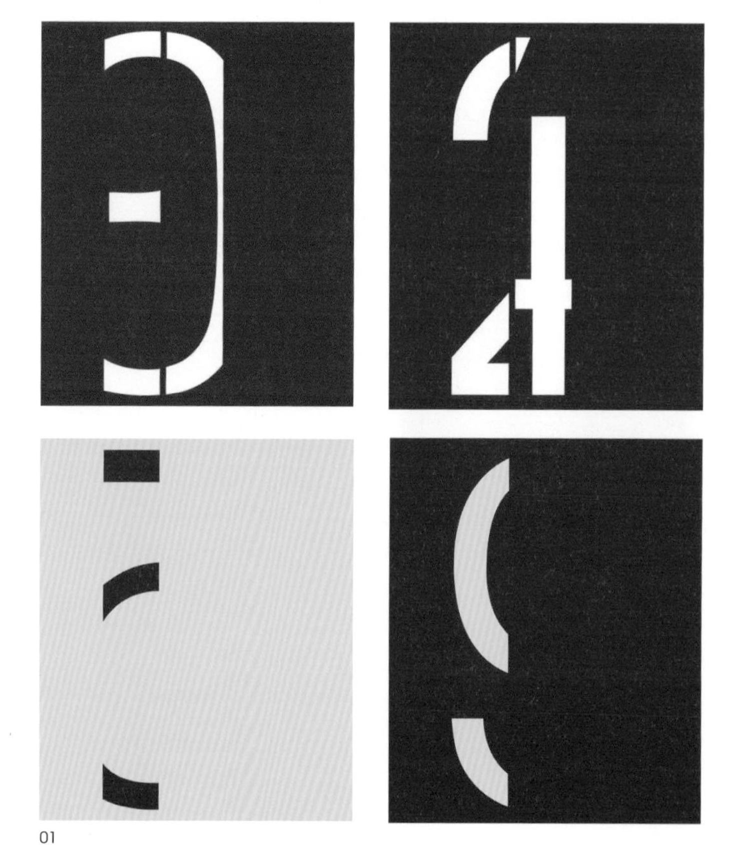

01

▸ 01

设计者：Struktur Design
作品名：*31 Days*
壁式年历

▸ 02

设计者：Struktur Design
作品名：*52 Weeks*
袖珍日记簿

November
05 Monday
06 Tuesday
07 Wednesday
08 Thursday
09 Friday
10 Saturday
11 Sunday

November
12 Monday
13 Tuesday
14 Wednesday
15 Thursday
16 Friday
17 Saturday
18 Sunday

November
19 Monday
20 Tuesday
21 Wednesday
22 Thursday
23 Friday
24 Saturday
25 Sunday

November | December
26 Monday
27 Tuesday
28 Wednesday
29 Thursday
30 Friday
01 Saturday
02 Sunday

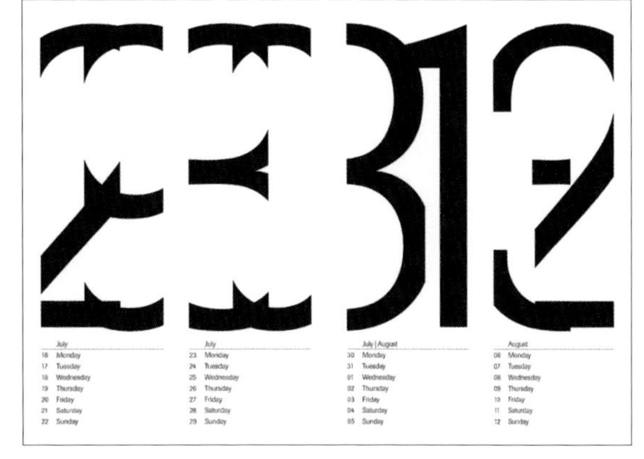

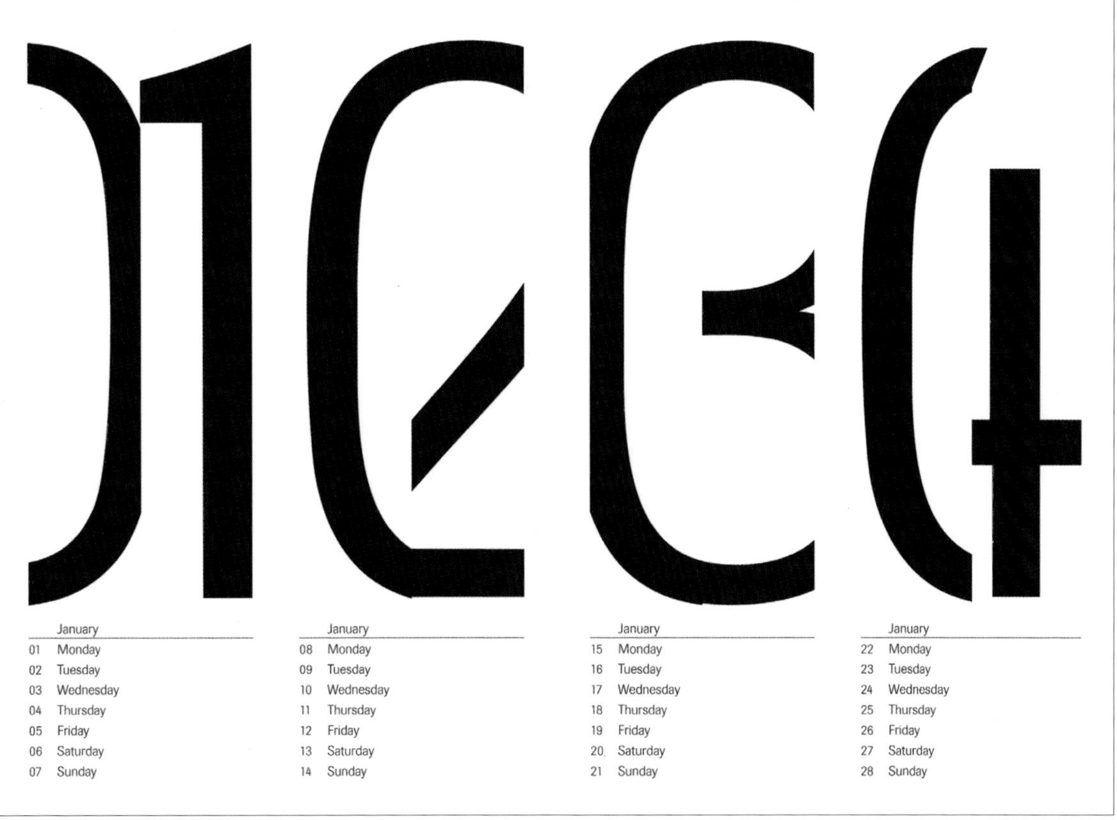

02

04.1.8
多项选择

《365》日历的每一页都有两处裂口，使用者可将页面分成3个部分。这款台历延续“斯图克尔”公司设计的日历系列，结合数字碎片来创造抽象的排版形式。可以通过许多种不同的方式来使用此日历，也可以将其当成一本传统的月历（1到12月）。

一年中365天的设计版式：左侧面板代表百位，中间面板代表十位，右侧面板代表个位。在适当的间隔后，左侧面板含有完整的日历详细说明并伴随可供选择的数字碎片。

当月的天数（1至31）：这种设计版式可以让左边的面板显示完整日历，日历显示的是3个月份的所有日期。

一年中的52周：这与之前的版式相似，但用的是两个较大的数字来表示一年中的52周。

01

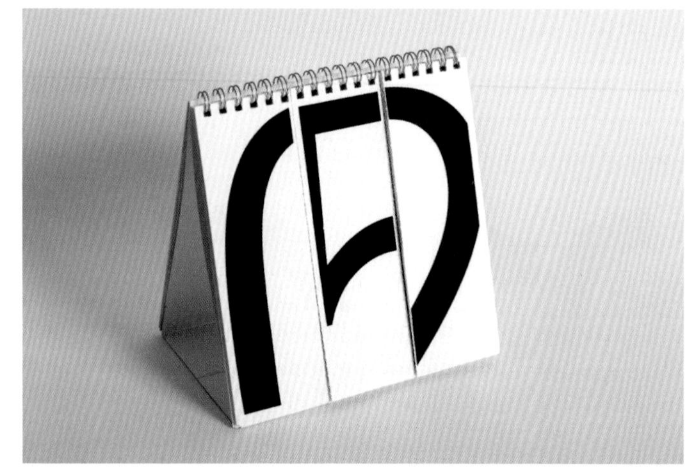

02

03

▸ 01

设计者：Struktur Design
作品名：*365*
该台历作为传统的月历使用

▸ 02

设计者：Struktur Design
作品名：*365*
该日历也可用来显示一年的第1至第52周

▸ 03

设计者：Struktur Design
作品名：*365*
该日历也可显示一年的第1至第365天

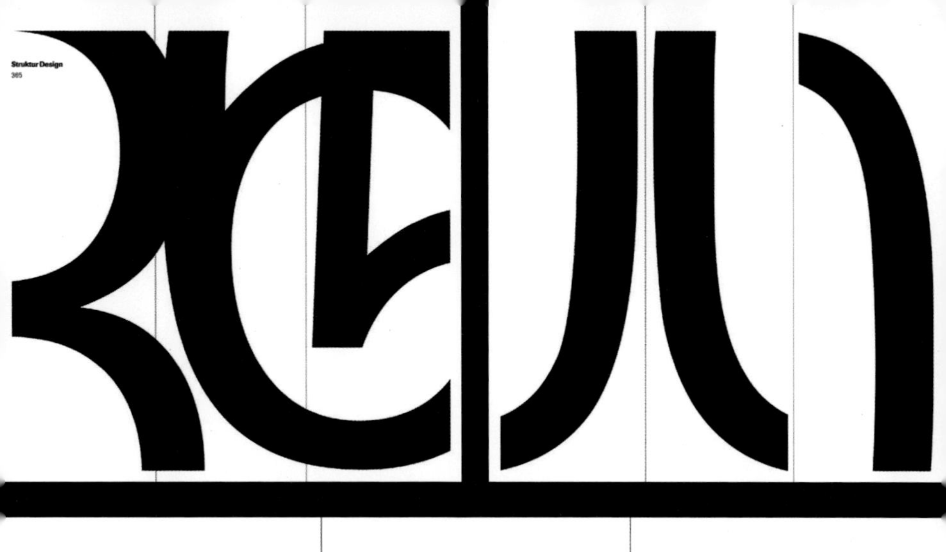

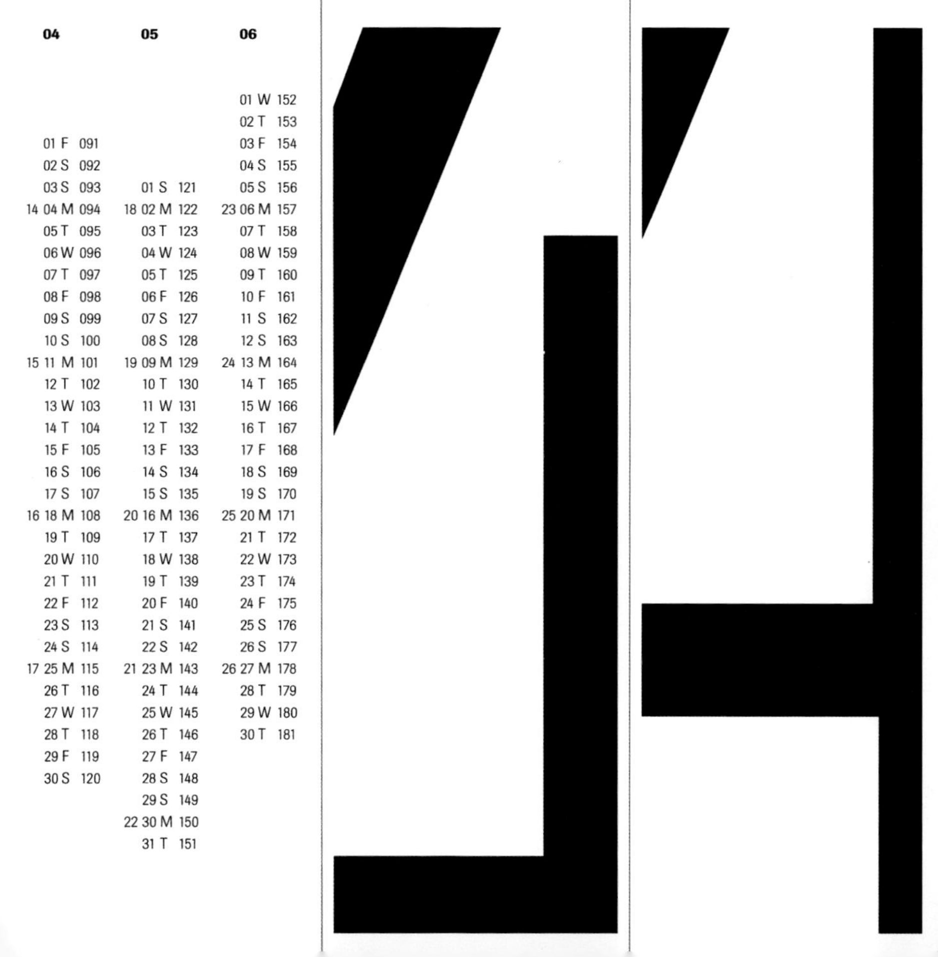

04

Week	Day		No.
	01	F	091
	02	S	092
	03	S	093
14	04	M	094
	05	T	095
	06	W	096
	07	T	097
	08	F	098
	09	S	099
	10	S	100
15	11	M	101
	12	T	102
	13	W	103
	14	T	104
	15	F	105
	16	S	106
	17	S	107
16	18	M	108
	19	T	109
	20	W	110
	21	T	111
	22	F	112
	23	S	113
	24	S	114
17	25	M	115
	26	T	116
	27	W	117
	28	T	118
	29	F	119
	30	S	120

05

Week	Day		No.
	01	S	121
18	02	M	122
	03	T	123
	04	W	124
	05	T	125
	06	F	126
	07	S	127
	08	S	128
19	09	M	129
	10	T	130
	11	W	131
	12	T	132
	13	F	133
	14	S	134
	15	S	135
20	16	M	136
	17	T	137
	18	W	138
	19	T	139
	20	F	140
	21	S	141
	22	S	142
21	23	M	143
	24	T	144
	25	W	145
	26	T	146
	27	F	147
	28	S	148
	29	S	149
22	30	M	150
	31	T	151

06

Week	Day		No.
	01	W	152
	02	T	153
	03	F	154
	04	S	155
	05	S	156
23	06	M	157
	07	T	158
	08	W	159
	09	T	160
	10	F	161
	11	S	162
	12	S	163
24	13	M	164
	14	T	165
	15	W	166
	16	T	167
	17	F	168
	18	S	169
	19	S	170
25	20	M	171
	21	T	172
	22	W	173
	23	T	174
	24	F	175
	25	S	176
	26	S	177
26	27	M	178
	28	T	179
	29	W	180
	30	T	181

04.1.9
挂历策划

“This”工作室设计的一系列年度挂历研究了组织年度日期的各种方法。《2008》采用简单的数字列表来呈现每一月份。

《2009》以景观版式将白色的嵌板印制于纯黑色的背景上，白色嵌板显示的是当前星期的每一天，周末则采用黑色。

《2010》采用放大日期的设计，丰富了传统的年历样式，为记载日记条目提供更多的空间。

《2011》重新定向数据：每月大量数字穿过海报，天数被写成单词而非数字。

《2012》重点在2011年的最后一天和2012年的第一天，2011年12月31日与2012年1月1日这两个日期占据海报的很大一部分。月份被排列成三部分，每部分列出4个月份。

01

02

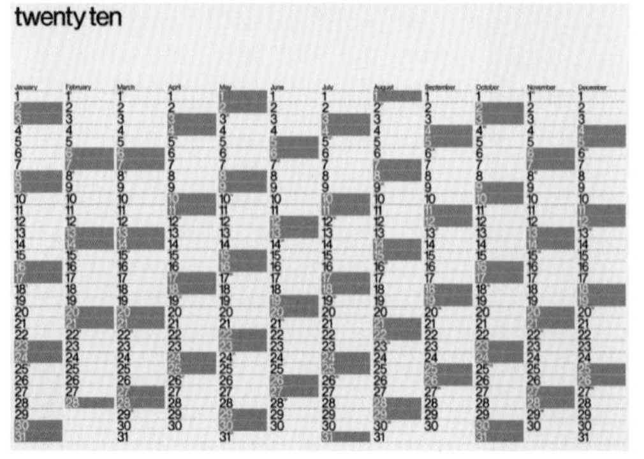

03

▸ 01

设计者：This Studio
作品名：*Two Thousand Eight*
挂历

▸ 02

设计者：This Studio
作品名：*Two Thousand and Nine*
挂历

▸ 03

设计者：This Studio
作品名：*Twenty Ten*
挂历

▸ 04

设计者：This Studio
作品名：*Twenty Eleven*
挂历

▸ 05

设计者：This Studio
作品名：*Twenty Twelve*
挂历

04

05

04.2
日记簿

04.2.1
年视图

瑞士编剧、作家马克思·弗里施（Max Frisch）诞辰100周年展览的目录可作为一本101年的日记簿，在日历中可查阅他的个人行程，该日历也对他工作和生活中的重要日期加以突出。每一页都会追溯十年期间与他生活对应时间发生的重大事件。该目录产生了一种有趣和吸引人的方案，可顺利地将其作为含大量信息的目录或时间轴，也可作为一本可用的日记簿，该日记簿同时含有弗里施的个人信息及对他本人来说非常重要的日期。

▸ 01
设计者：Eggers + Diaper
作品名：*Max Frisch 1911-2011*
展览目录与日记簿相结合

Wer sich nicht mit Politik befaßt, hat die politische Parteinahme, die er sich sparen möchte, bereits vollzogen: er dient der herrschenden Partei.

Das Atelier des verschollenen Stiller – (...) Wohin man blickte, hatte man in diesem Atelier das erregende Gefühl, jederzeit aufbrechen und ein ganz anderes Leben beginnen zu können, also genau das Gefühl, das Sibylle damals brauchte.

T : Homo faber | 208 Seiten | P
✻ : 1957
Walter Faber, Hanna Piper, Elisabeth (Sabeth), Ivy, Joachim Hencke
Mexiko, Guatemala, New York, Schiff, Frankreich, Italien, Griechenland, Havanna
Schicksal/Zufall, Schuld, Technikgläubigkeit, Versäumnisse, Mythologie, Inzest
! : Flugzeugabsturz, Schlangenbiss in Griechenland
☆ : Sam Shepard in Volker Schlöndorffs Film Homo faber 1991
märz
1910
1920
1930
1940
1950
Schülerzeitung «Mundus vult Schundus» kündigt Stück «Der Schüler im Himmel» von MF an

04.2.2
版式轮廓

“五角设计”工作室每年更新的《日历排版》中的每一月份均采用不同的字体，尽管每月遵循相同的平面排版，但由于字体的改变，每一月份看起来都与众不同。封面以数字“365”为特征，由12种古典和现代字体重叠而成，包括对其轮廓的设置。

“Base”设计公司为“La Casa Encendida”设计的日记簿，其特点是将最小限度的字体置于渐变的混合色之上，由深至浅然后又回到深色，以表现一年中的白天和黑夜。

“Remake”系列的学生手册与日历是为华盛顿特区的一家艺术与设计学院制作的。每年的学生手册以其信息设计的构造手法为特色，外加吸引人的影像——有时是版式、摄影或纯粹的图形。最终以迷人的方式展现出大量的不同内容。

01

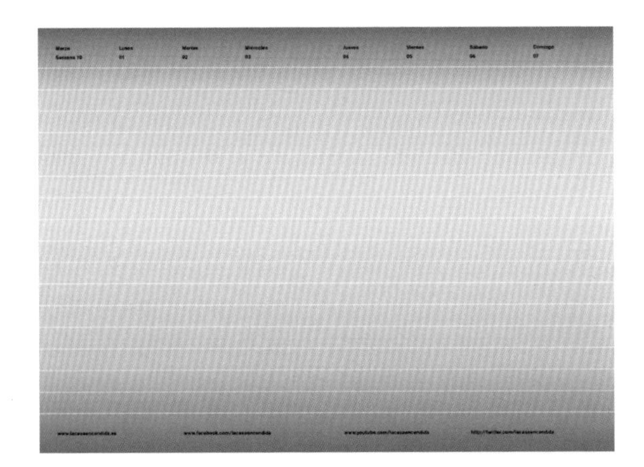

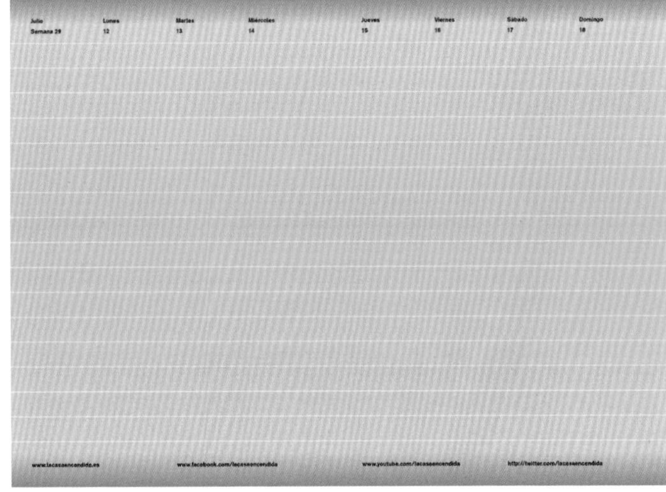

02

▸ 01

设计者：Pentagram
作品名：*2008 Typographic Calendar*
年度更新日历

▸ 02

设计者：Base Design
作品名：*La Casa Encendida diary*
日记簿

▸ 03

设计者：Remake
作品名：*Student Handbook and Calendar*
“Cororan”艺术与设计学院

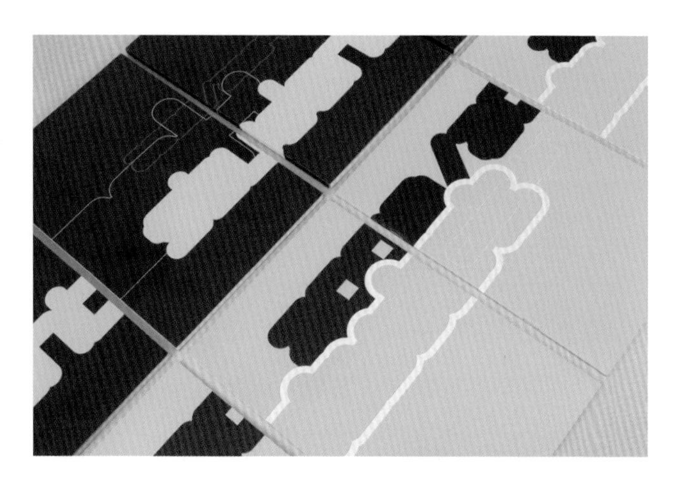

Corcoran College of Art + Design
2003/04 Student Handbook and Student Calendar

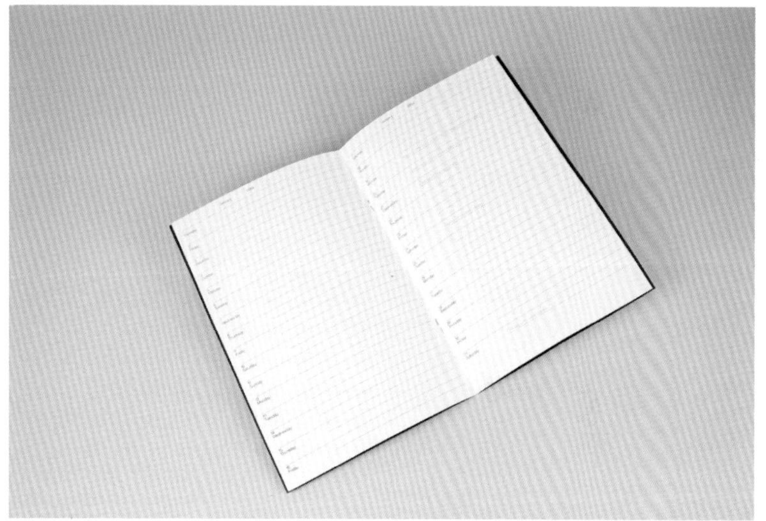

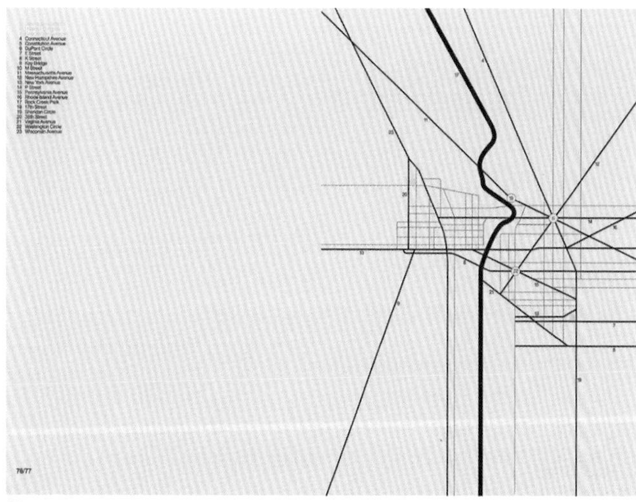

03

04.2.3
彩色序列

小型A6日记簿《时》的每次展开都代表了一个月份。左边的页面通过小时的累计总数来说明月份，从1月1日的第1个小时开始，到12月31日的第8 760个小时。每一月份中的小时均采用不同的颜色表示，月份中每一周的颜色逐渐加深。右手边的页面简单陈列出月份中的天数，印制成黑色，月份中的天数及年份中的周数用黄色表示。日记簿的最后，页面显示的是每个周末及公共假期中的小时数。

《信笺20》是由“SEA”设计公司的创始人之一自创的精品文具品牌。简洁、经典的排版设置与不断变化的背景颜色对应，呈现出季节变化。

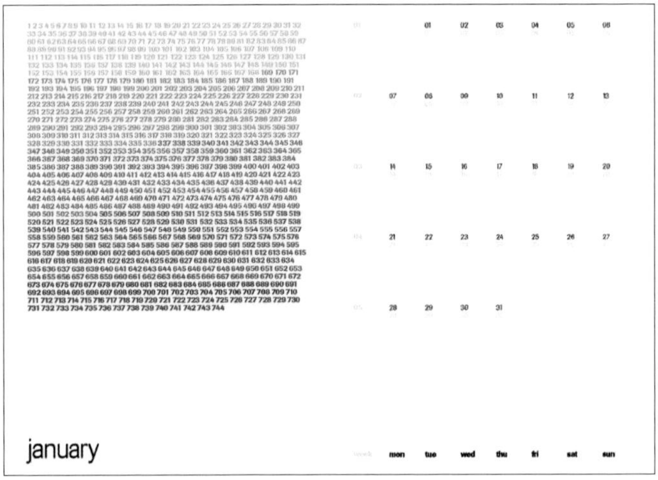

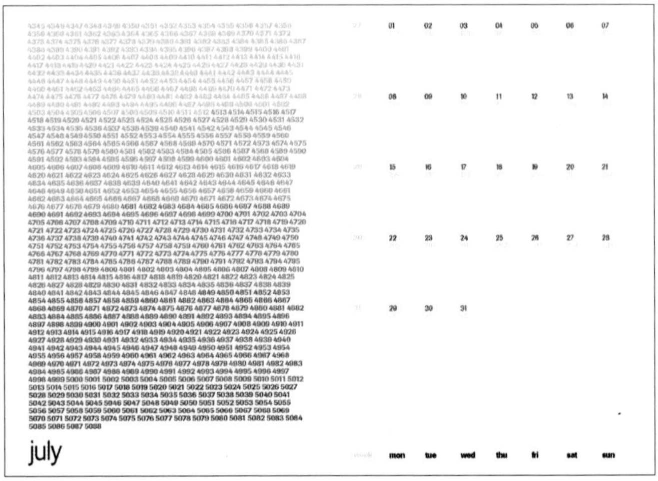

january

mon tue wed thu fri sat sun

july

mon tue wed thu fri sat sun

weekends

01

▶ 01

设计者：Struktur Design
作品名：*Hours*
A6日记簿

▶ 02

设计者：SEA Design
作品名：*20 Stationery*
办公日志

07
2007
01 Sunday
02 Monday
03 Tuesday
04 Wednesday
05 Thursday
06 Friday
07 Saturday
08 Sunday
09 Monday
10 Tuesday
11 Wednesday
12 Thursday
13 Friday
14 Saturday
15 Sunday
16 Monday
17 Tuesday
18 Wednesday
19 Thursday
20 Friday
21 Saturday
22 Sunday
23 Monday
24 Tuesday
25 Wednesday
26 Thursday
27 Friday
28 Saturday
29 Sunday
30 Monday
31 Tuesday
July
Juli
Juillet
Julio
Juli
Luglio

20
07

02

04.2.4
可用性

为“La Casa Encendida”设计的这一款日记簿，数字“2011”的碎片超出每页的边沿部分，这样设计的目的在于当日记簿合拢时可以显示出完整的年份。大面积的极小字体和白色空间与每一月份的开放式摆放形成对照，全部字体采用充满生机的绿色印制。日记簿只用一种颜色印制，但获得的效果却极富变化和趣味性。

▸ 01

设计者：Base Design
作品名：*La Case Encendida*
日记簿

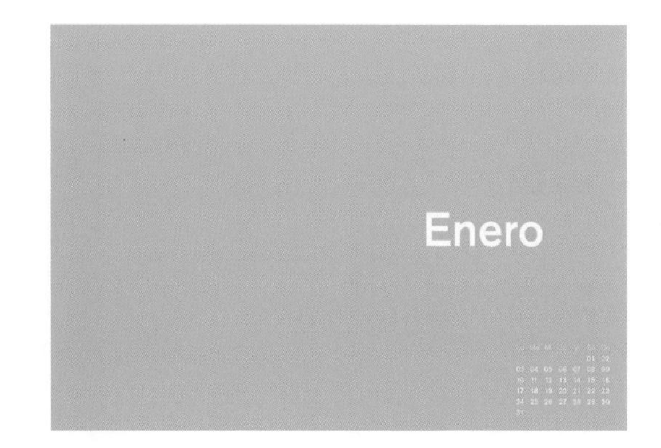
Enero

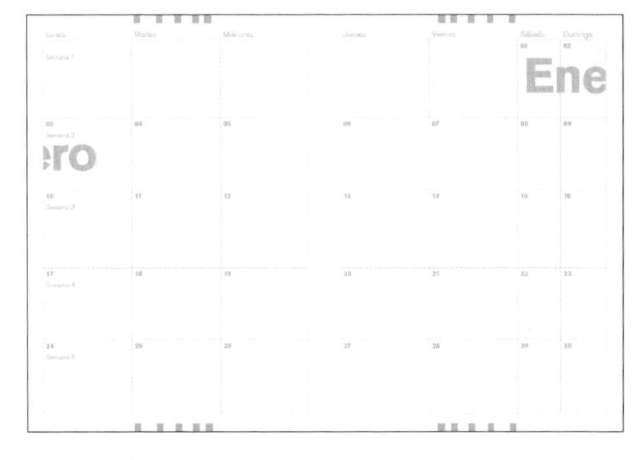
Ene
ero

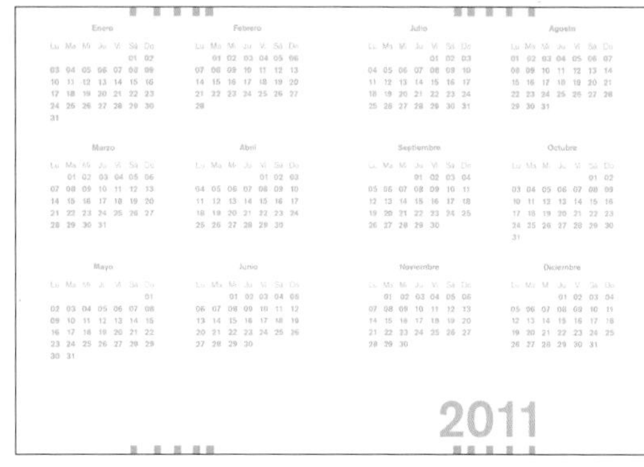
2011

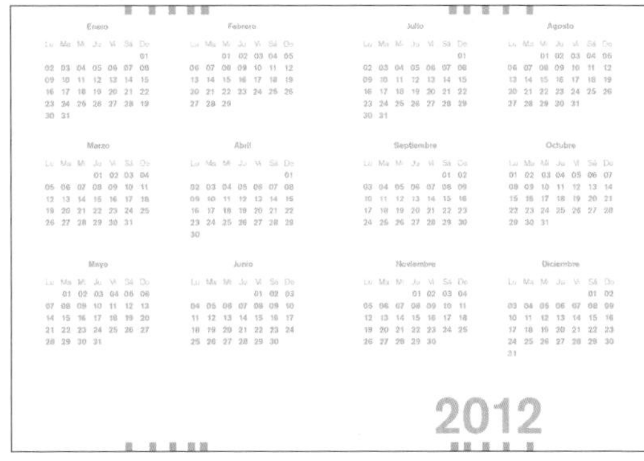
2012

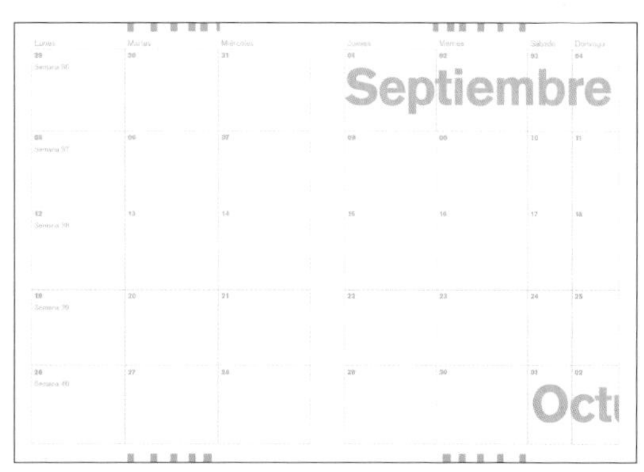
Septiembre

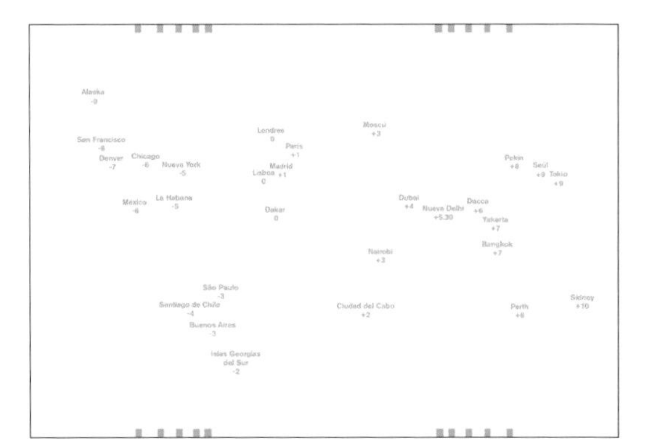
Alaska
San Francisco
Denver
Chicago
Nueva York
México
La Habana
Londres
París
Madrid
Lisboa
Dakar
Moscú
Dubai
Nueva Delhi
Dacca
Pekín
Seúl
Tokio
Bangkok
Nairobi
São Paulo
Santiago de Chile
Buenos Aires
Ciudad del Cabo
Perth
Sidney

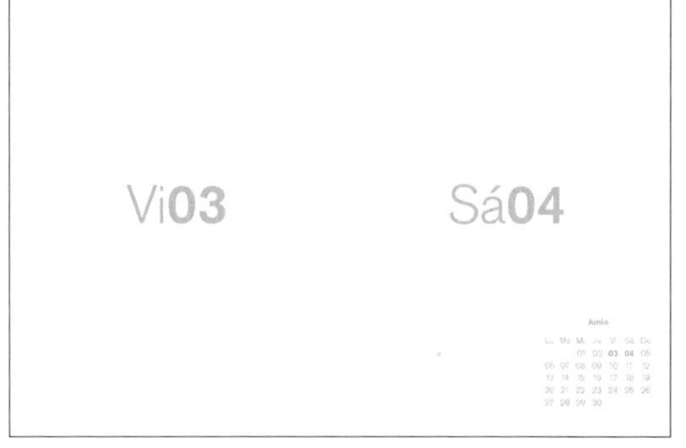
Vi03
Sá04

04.3
钟表

04.3.1
模拟数字

"Tonne"工作室设计的《24》是代表24小时时光流逝的出色例子。从左下角00:00开始，每分钟在嘀嗒声中向右前行。时间的角度直接模拟指针式时钟上分针的角度，所以00:00的读数为垂直的，00:15为水平的，而00:45则为倒置的。由此通过字体角度的变化创造出了优美的节奏曲线。

《诞生时钟》是一份庆祝罗莎出生的私人礼物，时间指向11：24。.

01

01

设计者：Studio Tonne
作品名：*24*
按钟的指针所指时间进行旋转的分钟数字来表示分钟

02

设计者：Studio Tonne
作品名：*Birth Clock*
永远停留的一瞬间

02

04.3.2
数码数字

由斯洛伐克图形与字体设计师“Ondrej Job”设计的“ION”字体总共有三类字体家族：浓缩型（ION C）、标准型（ION B）及宽广型（ION C）。图形以经典的七段式LCD显示为基础。每种“ION”字体家族均具有引人注目的十种粗细级别，可支持70余种拉丁系语言和各种开放型功能，包括自由连字、分数及语体设置。该字体还具有一种被称为“Cells”的特殊框绘字体。

《时间：音调》是为“Mute”唱片公司设计的一款音频时钟和新闻订阅程序。基于对测量的着迷，这一可下载信息的小物件可以根据设定的间隔时间来播放音乐（来自“Mute”唱片大量的备份目录）：针对年、月、日、小时、分钟、秒钟等播放不同的音乐。也可以自动下载新闻、发布以及直播“Mute”固定艺术家的事件信息。

“马尔科姆·韦伯”（Malcolm Webb）是来自“Tonne”工作室的一项个人项目，为庆祝马尔科姆·韦伯40岁生日而设计。每一字符均由布局在10×4网格中的数字“40”构成；字符由黑色和灰色的数字“40”结合而成。年代数被分解成周、日、时、分和秒。

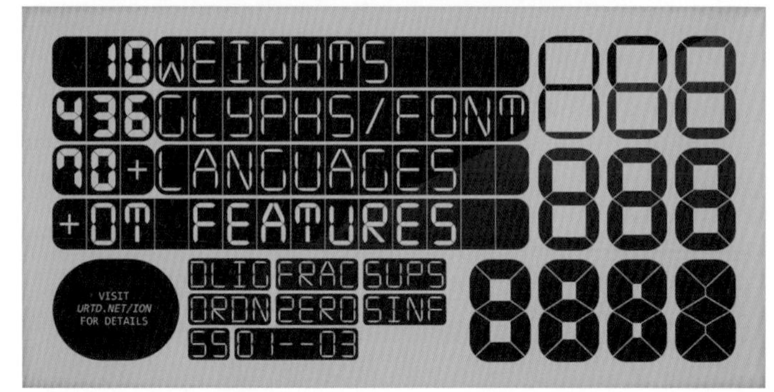

01

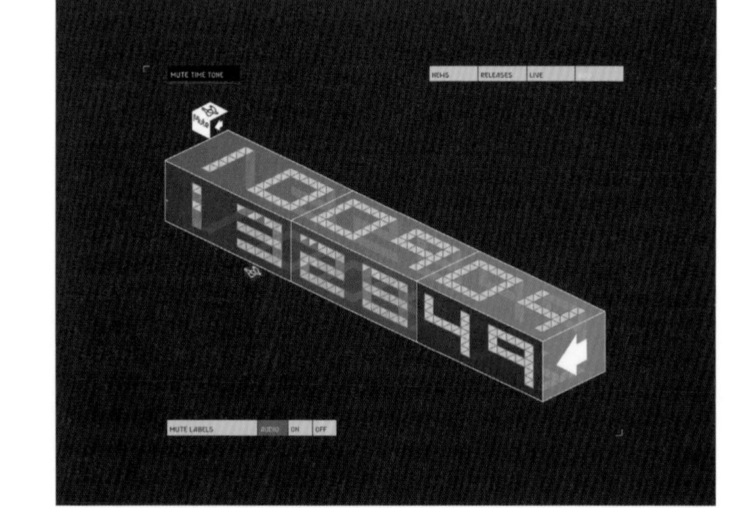

02

▶ 01

设计者：Ondrej Job
作品名：*urtd. Net*
ION
字体设计

▶ 02

设计者：Studio tone
作品名：*Time: Tone*
“Mute”唱片公司的音频时钟及新闻应用程序

▶ 03

设计者：Studio tone
作品名：*Malcolm Webb 40th birthday greeting*

1,262,304,000 SECONDS

21,038,400 MINUTES

350,640 HOURS

14,610 DAYS

2087 WEEKS

40 YEARS

MALCOLM WEBB
26 JANUARY 1971 9.45PM

HAPPY BIRTHDAY
FROM THE FARRINGTONS

04.3.3
原子钟

"Tonne"工作室受委托为奥迪客户杂志设计一篇文章。该杂志的此次发行以24小时为主题，《无论24小时有多长》提供了世界各地230个原子钟如何同步的视图，通过无线电信号连接至时钟和计算机网络。

每一秒钟被定义为一个铯-133原子的9 192 631 770次振动。科学家对原子核周围单电子的波动进行了测量。总共的电子数为55——其中有54个位于原子核周围，另一个位于环绕其他电子的"轨道"。

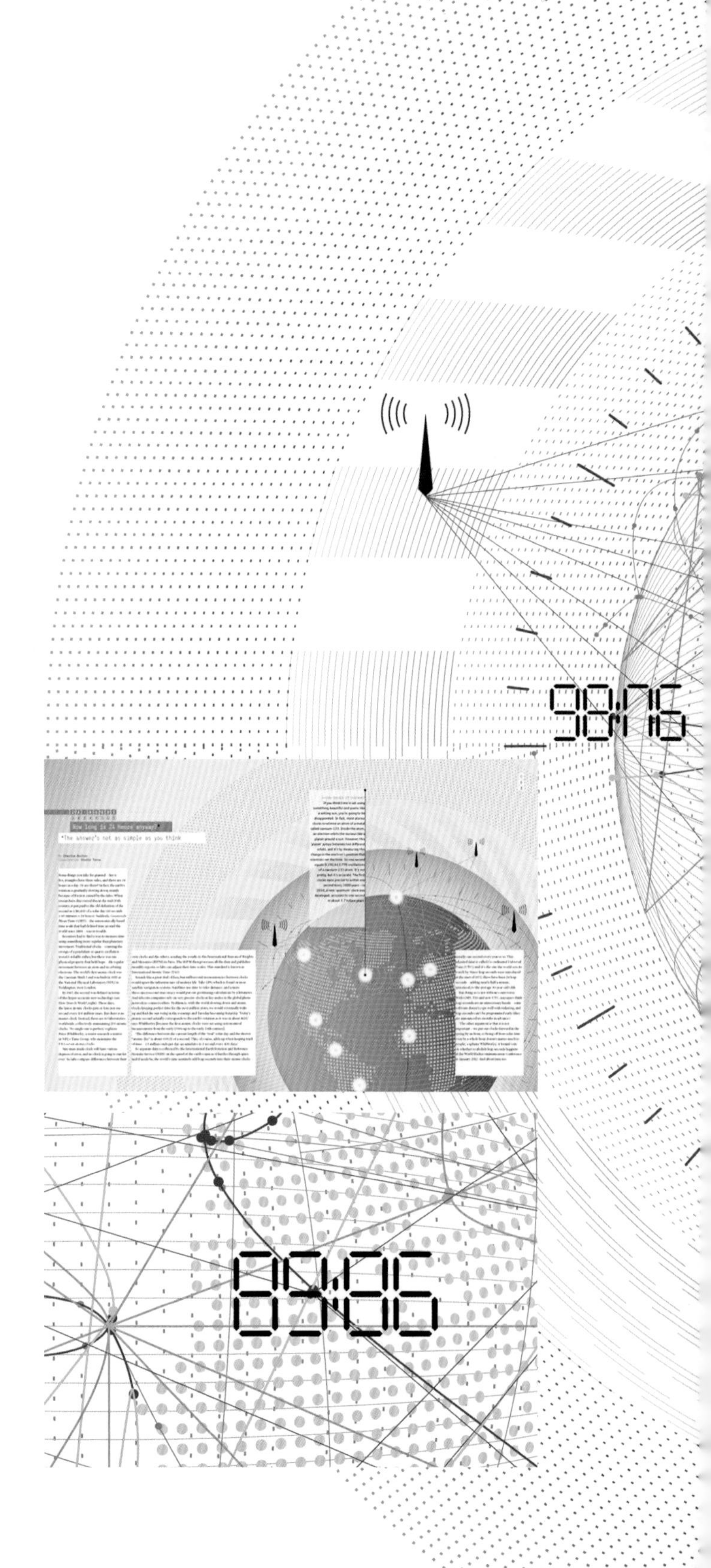

▸ 01

设计者：Studio Tonne
作品名：*How Long are 24 Hours Anyway?*
奥迪杂志

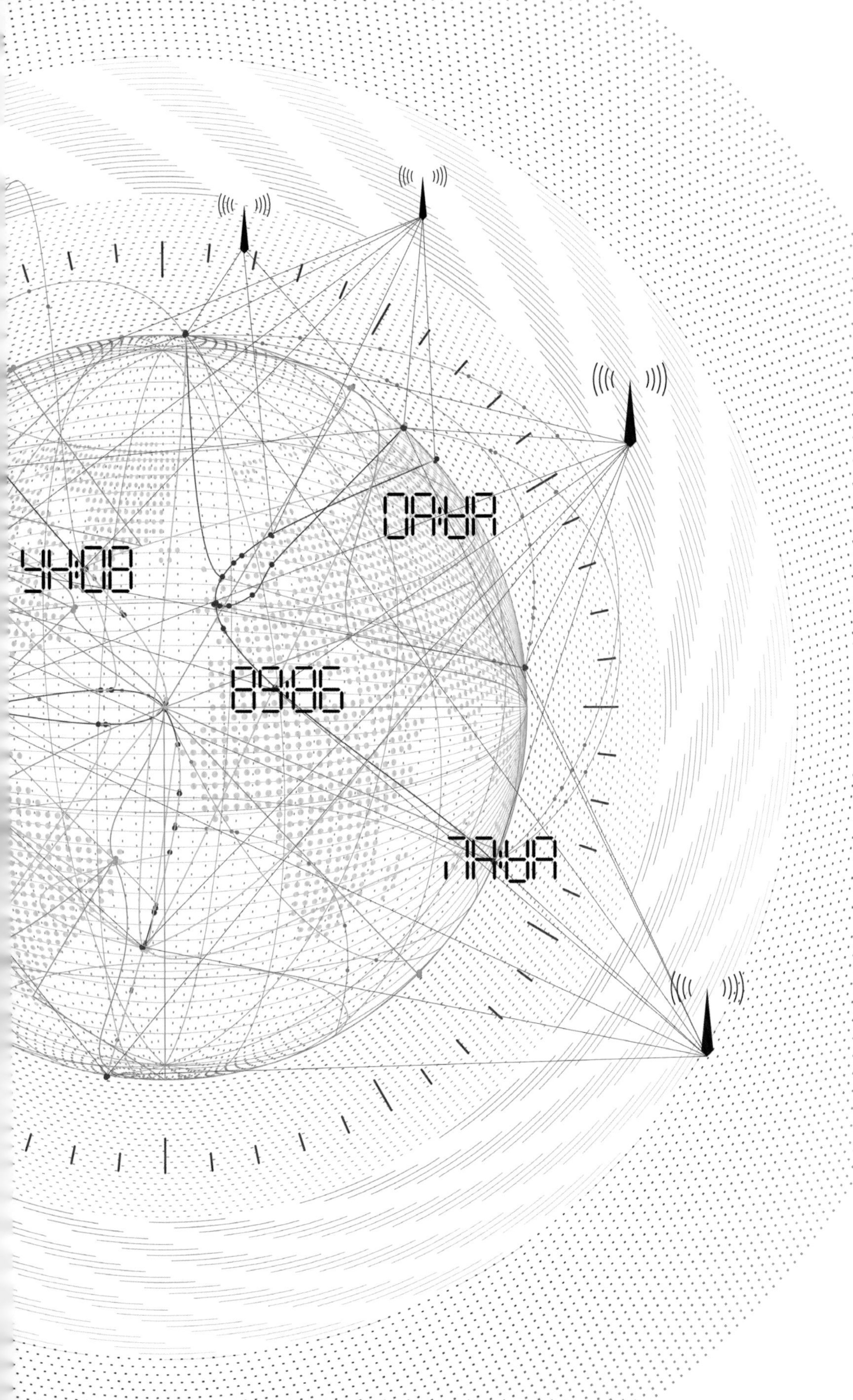

04.3.4
时钟应用程序

《时间设备01》是一款苹果手机应用程序，由“琦莉·X”（Chilli X）与字体设计师瑞恩·休斯（Rian Hughes）联合设计。时与分以黑色粗体字显示，秒的颜色在一分钟内逐渐变深。背景颜色可以自动变化，它会通过光谱逐渐消退，也可以通过手动——摇晃手机来随机选定背景颜色。

《NightTime plus》是一款苹果手机闹钟，其设计灵感来自传统的LCD显示屏。该显示屏可以自定义不同的颜色和纹理。

《关于时间》诉说了这是什么时间，产品特色是在浅色羊皮纸背景上呈现的漂亮字体。在白天为亮色而在晚上则是深色。它创造出的不仅仅是时钟，而是一种理念（不同寻常的）。不但以一种独特和非传统的方式报时，还可以快速翻动页面来显示关于时间本身有趣的语录及感言。

《F-Clock》是一款分片式时钟，将模拟概念与数字旋转结合。《U-Clock》以一系列时间栏来显示在屏幕上移动的时间。

01

02

03

04

05

06

▸ 01
▸ 02

设计者：Chilli X and Rian Hughes
作品名：*Time Device 01*
苹果手机时钟

▸ 03

设计者：Chilli X
作品名：*Night Time*
苹果手机时钟

▸ 04

设计者：Chilli X
作品名：*Night Time Plus*
苹果手机时钟

▸ 05
▸ 06

设计者：Chilli X
作品名：*About Time*
苹果手机时钟

▸ 07
▸ 08

设计者：Winfield & Co
作品名：*F-Clock*
苹果手机时钟

▸ 09
▸ 10
▸ 11

设计者：Winfield & Co
作品名：*U-Clock*
苹果手机时钟

PM
1257

07

PM
1234
MON
JAN
30
2012

08

HOURS
MINUTES
SECONDS
1/100 SECONDS

09

HOURS
MINUTES
SECONDS
1/100 SECONDS

10

HOURS
MINUTES
SECONDS
1/100 SECONDS

11

04.3.4
时钟应用程序

《文字时钟》是“Commune Inc.”的一款苹果手机应用程序，将时间看作是词语序列——无任何数字。

《嘀嗒时钟》以指针式电子钟的圆周运动为基础。分钟数及秒数绕着位于屏幕中心的小时数运行。

《单词时钟》是一款针对苹果电脑、平板和手机设计的版式时钟和交互式工艺品。通过强调突出单词（30多种语言中的一种）来显示时间。用户可以选择线型或动态的旋转版本，或在两种版本间定期转换。

12

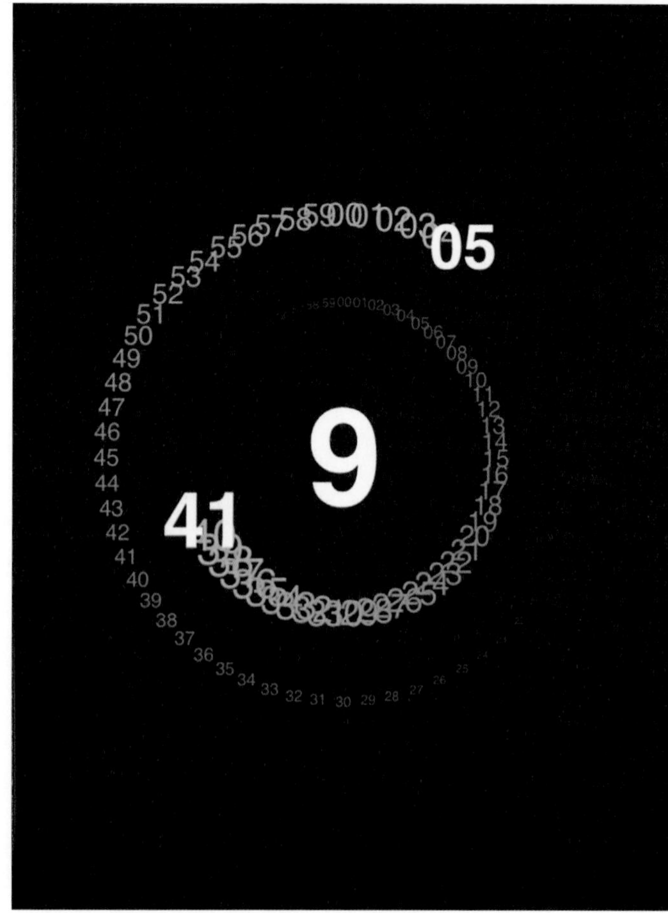

13

▸ 12

设计者：Commune Inc
作品名：*Text Clock*
苹果手机时钟

▸ 13

设计者：Coffeecoding
作品名：*ClickClock*
苹果手机时钟

▸ 14

设计者：Simon Heys
作品名：*WordClock*
苹果电脑、平板和手机的屏幕保护程序

One Two Three Four Five Six Seven Eight Nine Ten Eleven Twelve o'clock oh-one oh-two oh-three oh-four oh-five oh-six oh-seven oh-eight oh-nine ten eleven twelve thirteen fourteen fifteen sixteen seventeen eighteen nineteen twenty twenty-one twenty-two twenty-three twenty-four **twenty-five** twenty-six twenty-seven twenty-eight twenty-nine thirty thirty-one thirty-two thirty-three thirty-four thirty-five thirty-six thirty-seven thirty-eight thirty-nine forty forty-one forty-two forty-three forty-four forty-five forty-six forty-seven forty-eight forty-nine fifty fifty-one fifty-two fifty-three fifty-four fifty-five fifty-six fifty-seven fifty-eight fifty-nine **and** precisely one second two three four five six seven eight nine ten eleven twelve thirteen fourteen fifteen sixteen seventeen eighteen nineteen twenty twenty-one twenty-two twenty-three twenty-four twenty-five twenty-six twenty-seven twenty-eight twenty-nine thirty thirty-one thirty-two thirty-three thirty-four thirty-five thirty-six **thirty-seven** thirty-eight thirty-nine forty forty-one forty-two forty-three forty-four forty-five forty-six forty-seven forty-eight forty-nine fifty fifty-one fifty-two fifty-three fifty-four fifty-five fifty-six fifty-seven fifty-eight fifty-nine **seconds**

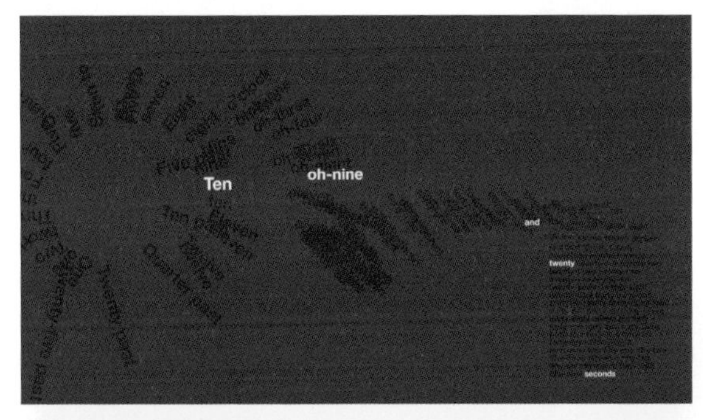

Twenty-five past
six
and
thirty-one
seconds

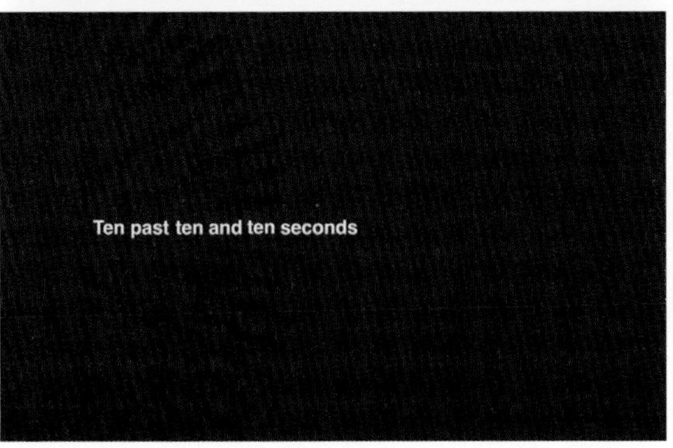

14

04.3.5
探索时间

"SEA"设计公司向《ICON》杂志提出了一系列手表界面更新的提案。这些设计方案呈现出对21世纪数字显示手表的重新思考，摆脱了传统LCD显示的限制。

"法罗"设计（Farrow Design）受现代家具公司"SCP"委托，设计了一系列壁挂钟。《Notime，Nightime and Finetime》以传统的指针式电子钟为基础，以简洁和精致的图案为特点，呈现出全新的现代风格。

01

▸ 01
设计者：SEA Design
作品名：*Icon Watches*
手表概念设计

▸ 02
设计者：Farrow Design
作品名：*Notime*
SCP壁挂钟

▸ 03
设计者：Farrow Design
作品名：*Nightime*
SCP壁挂钟

▸ 04
设计者：Farrow Design
作品名：*Finetime*
SCP壁挂钟

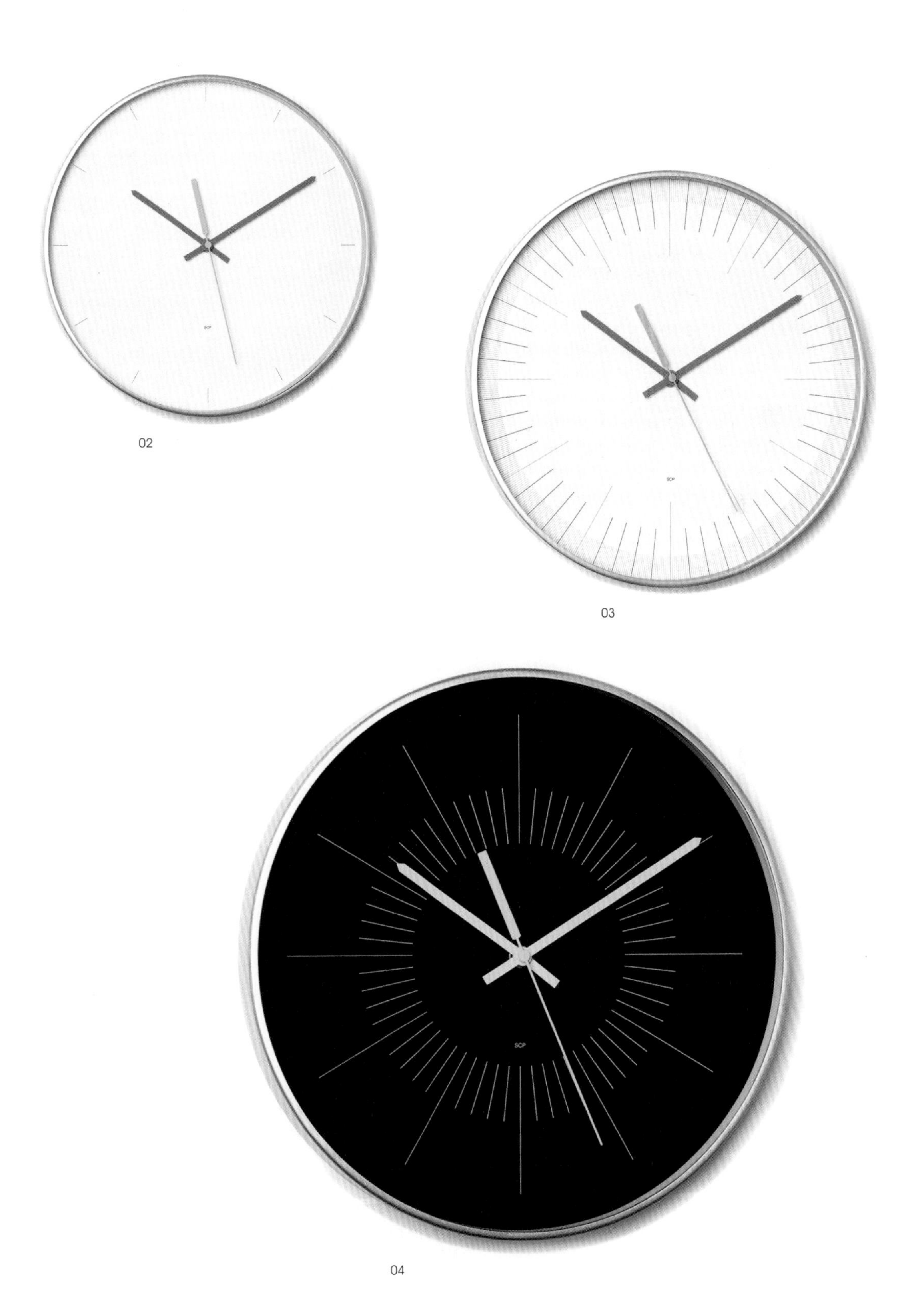

02

03

04

04.3.6
组织时间

“斯图克尔”公司设计的“斯图克尔时钟”重组并合理化了传统的指针式时钟。以24小时时钟为基础，编排好的小时及分钟位于方形时钟正面两侧的边缘。

《24小时时钟》是由总部位于伦敦的专业印刷商“Artomatic”制作的一种开放式简报。真丝丝网印刷的A2海报是作为着眼于时间系统的某一正在进行的研究项目的一部分而设计的，其是“斯图克尔”公司日历与日记簿的直接延伸（见第128至135页，142页）。海报陈列出24小时时段内的每一秒钟、每一分钟和每一小时，每一时间计量单位以逐渐变大的形式重现。

01

▸ 01

设计者：Struktur Design
作品名：*Struktured Clock*
指针式时钟

▸ 02

设计者：Struktur Design
作品名：*Twenty Four Hour Clock*
A2限量版海报

An active termite colony can eat
an entire house in just **3** years.

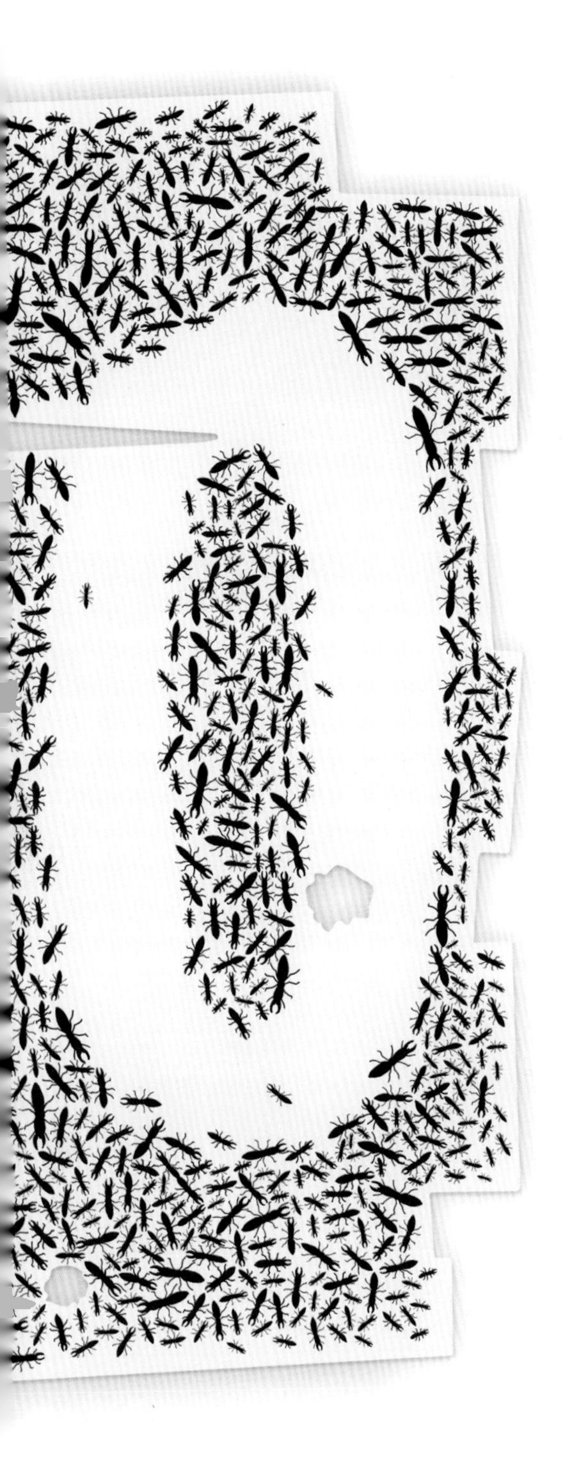

05.
抽象化

05.1 变形

05.1.1 立体形式 162
05.1.2 数字的自发形成 164
05.1.3 手绘 168
05.1.4 圆和线条 170
05.1.5 数字插画 172
05.1.6 将字体当作图形 174

05.2 地方的

05.2.1 遗失与发现 176
05.2.2 图形元素 180
05.2.3 空间里的数字 182
05.2.4 划痕与燃烧 184
05.2.5 数字无处不在 186
05.2.6 现代与古典 188
05.2.7 极简主义 190
05.2.8 超文本设计 192

05.3 字体设计

05.3.1 组合形式 196
05.3.2 抽象体 198
05.3.3 源于字母的数字 200

05.1
变形

05.1.1
立体形式

《Focus Open 2011》国际设计大奖的宣传单以大尺度立体文字为特点，赋予其空间感。“2011”年度这一数字随意置于单词“open”的旁边。数字“0”可与“open”中的“o”互换。书的封面由这些立体形式与年度主打产品一起构成，将字母转换为一件件家具。

《Firestation Workshops 2004》海报，来自都柏林的一家艺术家制片中心，在排版上使用一种漏印字体，分散于整个页面。数字“2004”采用红色印制，与并列式显示形成对照，同时数字“0”与海报右下部的雕塑形式相呼应。

01

02

▸ 01

设计者：Stapelberg & Fritz
作品名：*Focus Open 2011*
巴登符腾堡州
国际设计大奖宣传单

▸ 02

设计者：Stapelberg & Fritz
作品名：*Focus Open 2011*
巴登符腾堡州
国际设计大奖年度图书封面

▸ 03

设计者：Atelier David Smith
作品名：*Firestation Workshops 2004*
工作室活动海报

FIRESTATION

WORKSHOPS

2004

John Carrick / Rita Duffy / Noel Kelly / David Kinnane / Caroline Madden / Deirdre McLoughlin / Tanya Elliot Nyegaard / Geraldine O'Neill / Niall O'Neill / Alan Phelan

The annual artists' workshop programme offers exciting, challenging and innovative opportunities for the career development of visual artists

Fire Station Artists' Studios, 9/11 Buckingham Street, Dublin 1. Telephone: 01 855 6735. Fax: 01 855 5632

www.firestation.ie

03

05.1.2
数字的自发形成

“五角设计”自创的海报《数字系列》对易读性的界限进行了探索。每一款海报在文体上都与下一款截然不同。但是黑色的连续使用使得这些系列连在一起时，只要瞥一眼就可以看出其他的颜色。

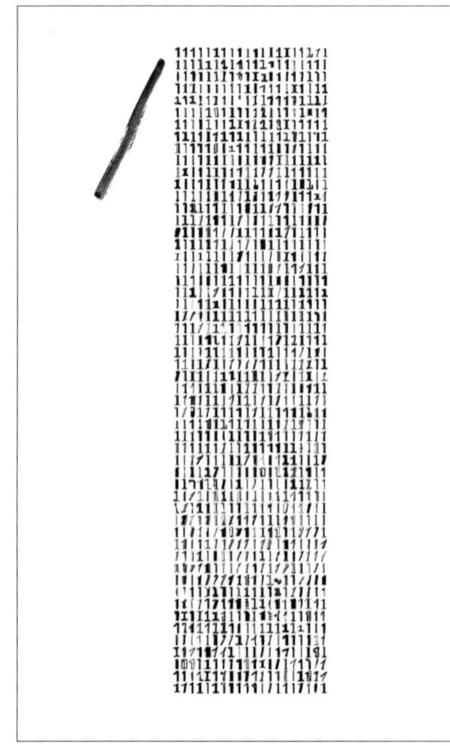

01
设计者：Pentagram
作品名：*The Number Series*
海报0–5

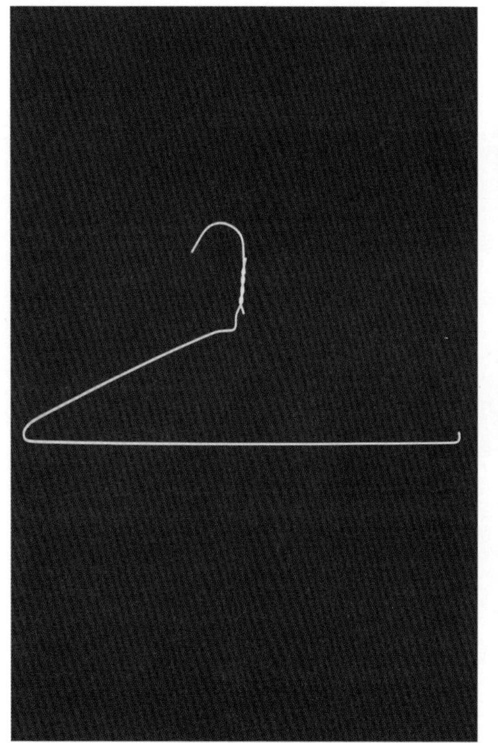

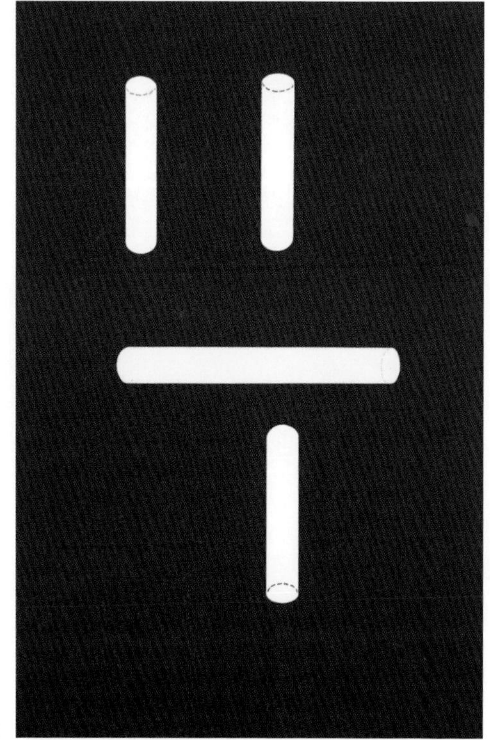

01

05.1.2
数字的自发形成

这些海报来自“五角设计”的《数字系列》，对负空间的使用进行了探索；数字“6”由一个长方形和一个正方形简单生成。以灰色背景上的两个白色三角形来突出数字“7”，数字“8”转变成两个椭圆，数字“9”由两轮新月与一个圆组成。

02
设计者：Pentagram
作品名：*Posters 6-9*
海报6–9

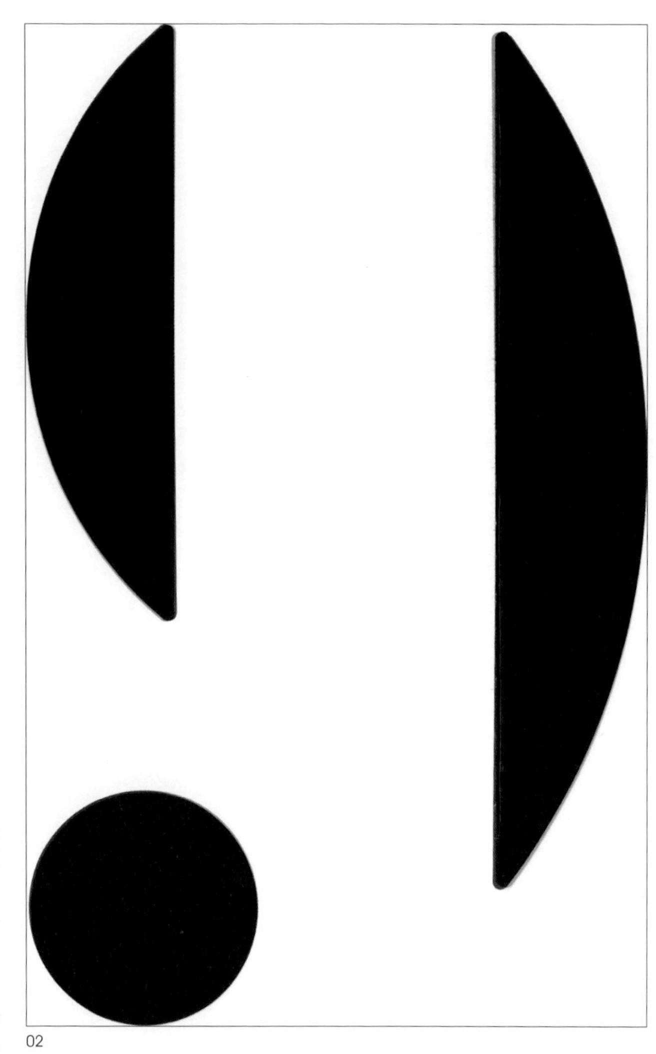

02

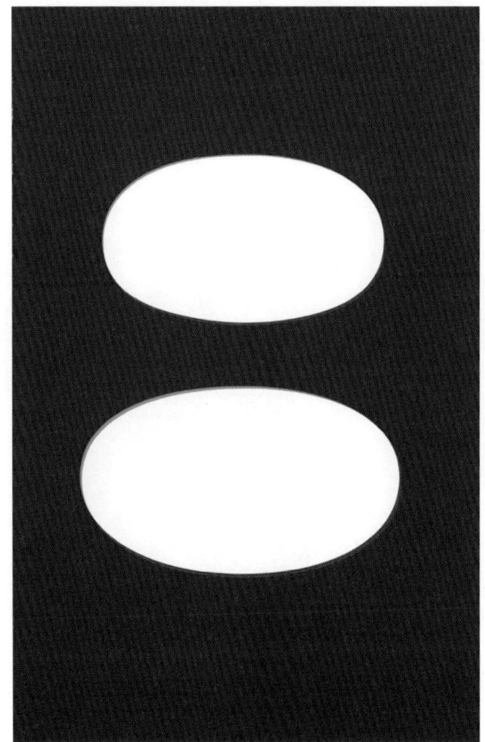

05.1.3
手绘

《米妮的八岁生日》是设计师的女儿8岁生日派对的邀请卡，派对在当地一个溜冰场举行。米妮用不同颜色的LED灯绘成数字“8”。

“邦奇”设计（Bunch Design）为迪赛的一个时尚品牌“55DSL”设计的宣传资料，对数字“55”和“555”以手绘图形的方式进行处理。“55DSL”大型女用手提包的特点在于数字“55”以交错的绣花刺绣条构成，且印上了公司的企业特有颜色：红、白、蓝。

《555》海报以圣诞节期间的一项店内活动——“55DSL你的圣诞节都录音了吗”为基础。为此鼓励员工将特别印制的绿色包装带应用于店内的几乎所有静物，而海报也通过用绿色胶带裹住三个数字“5”来延用这一主题。

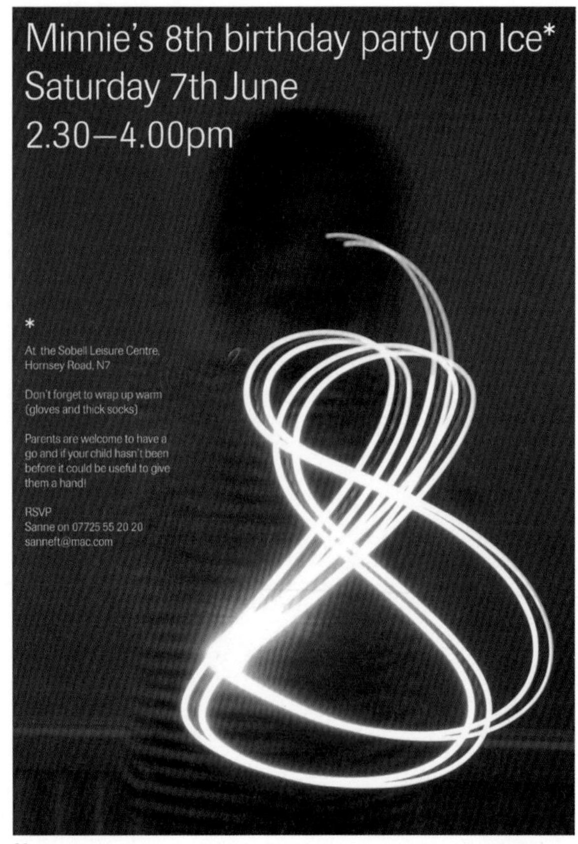

01

02

▸ 01

设计者：Struktur Design
作品名：*Minnie's 8th Birthday*
光绘的生日邀请卡

▸ 02

设计者：Bunch Design
作品名：*55DSL Tote Bag*
限量版丝网印刷包

▸ 03

设计者：Bunch Design
作品名：*555 Poster*
“55DSL”圣诞期间店内促销活动

03

05.1.4
圆和线条

《GP is 10》是印刷商“Generation Press”的一款两用限量版宣传手册，既用于庆祝他们的10周年纪念日，也向新客户和老客户展示他们印刷的超凡能力和技术。封面的特点在于标题采用蓝绿色、洋红及黄色进行套印，数字“10”采用黑色加以重叠。内部为呈现不同印刷流程的一系列折叠式纸片。包含的是主要由圆形元素构成的一种有趣的数字型设计。

01
设计者：Build
作品名：*GP is 10*
宣传手册

05.1.5
数字插画

儿童图书《数字的虫子》是充满虫类图形和数字的有趣教育资源。每一个展开的页面关注一种不同的昆虫，完全通过数字不同的粗细、大小和字体来说明。该图书还包括模切图形和原图，用来进一步激起年轻读者的兴趣。

01
设计者：Werner Design Werks
作品名：*Bugs the Numbers*
儿童书籍

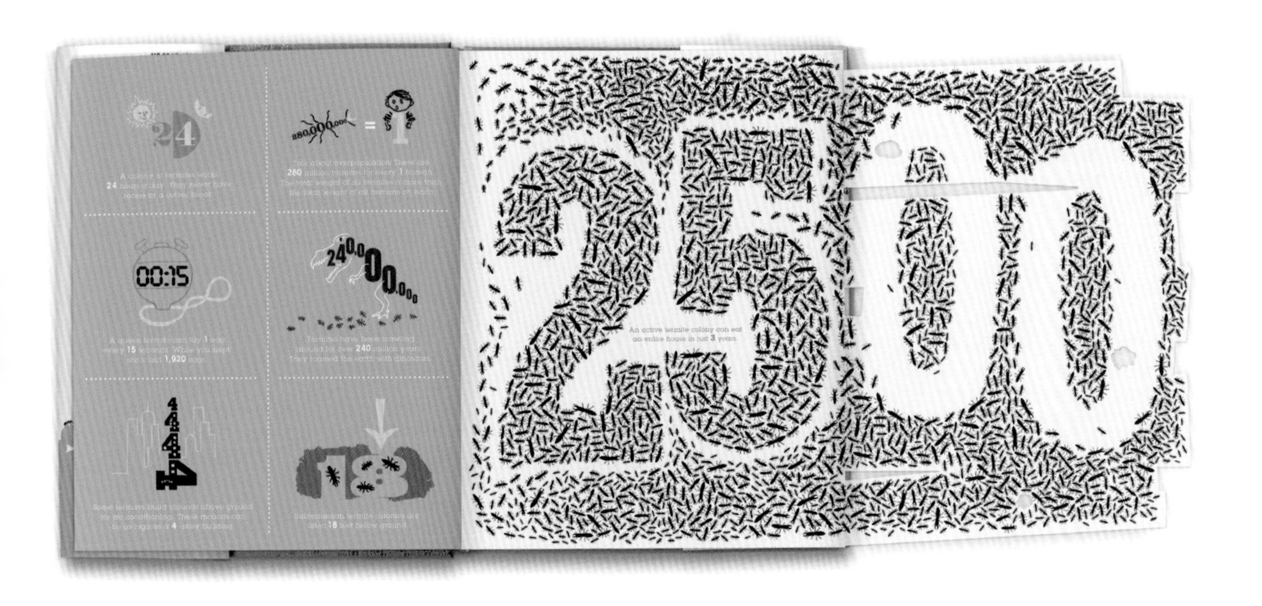

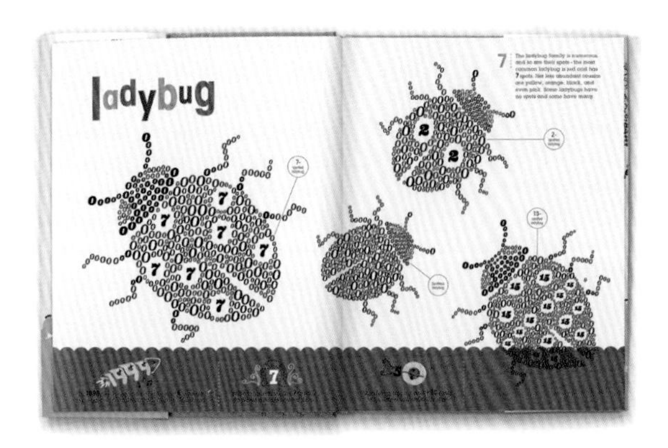

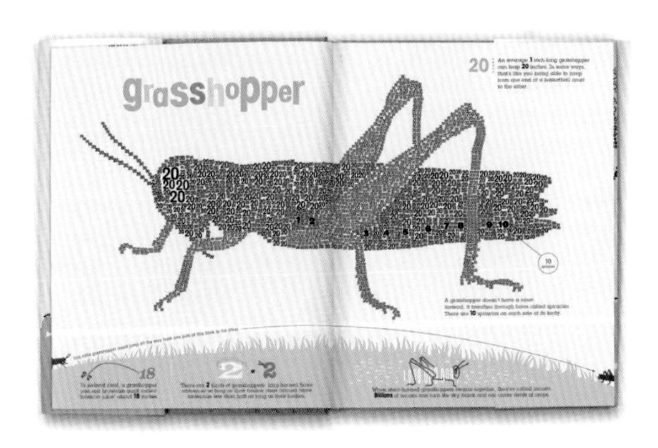

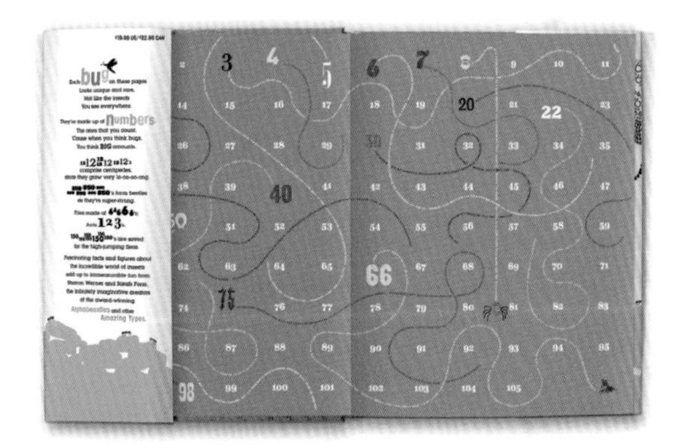

05.1.6 将字体当作图形

"Werner Design Werks"连续8年进行《WDW数字玻璃杯》的制作，《WDW数字玻璃杯》是一系列赠送给朋友、客户及供应商的鸡尾酒杯。每年的数字和鸡尾酒配方均不同。其理念是：保持简单、实用并保证配方口味良好。历年来对字体的巧妙选择成为有趣混合版式的一部分。

"邦奇"设计（Bunch Design）为"Consultants Group"设计的版式台历上的每一月份通过数字插图来印制，比如二月由"2"组成的天鹅心来表示，七月由"7"组成的冰淇淋来表示。

"Erotipo"是指版式设计的"色情性"，它是莱奥纳多・索罗里（Leonardo Sonnoli）为在威尼斯建筑大学开设的一项课程设计的主题。该课程是索罗里与艾琳・巴基（Irene Bacchi）对字体、排版及人体之间关系研究的结果。这里列出的实例来自索罗里在周刊《FF X-TRA》上发表的八项有关人类形体的问题。

01

02

▸ 01

设计者：Werner Design Werks
作品名：*WDW Number Glasses*
促销鸡尾酒杯

▸ 02

设计者：Bunch Design
作品名：*Typographic Desk Calendar*
"Consultants Group"的促销台历

▸ 03

设计者：Leonardo Sonnoli
作品名：*FF X-TRA*
带有色情味的版式设计试验

03

05.2
地方的

05.2.1
遗失与发现

莱昂纳多·索罗里（Leonardo Sonnoli）、莎拉·方纳利（Sara Fanelli）及乔治·哈迪（George Hardie）合作为第四届《如果你能合作》展会的一系列数字进行设计，该展会是伦敦艺术导演威尔·哈德森（Will Hudson）和艾力克斯·贝克（Alex Bec）自创的项目。

展会的参与者要求与来自任何学科、专业或背景的伙伴联合制作一些出人意料的作品，这些作品没有答案的限制或遵循的格式。本次展会在伦敦罗谢尔学院的基金会画廊上对来自33个团队的作品进行展示。展品的类目从经典框架类作品到更宏伟、更具试验性的雕塑、影像及设施。

一系列数字类型：从手绘涂鸦到具有地方特色的不同字体。

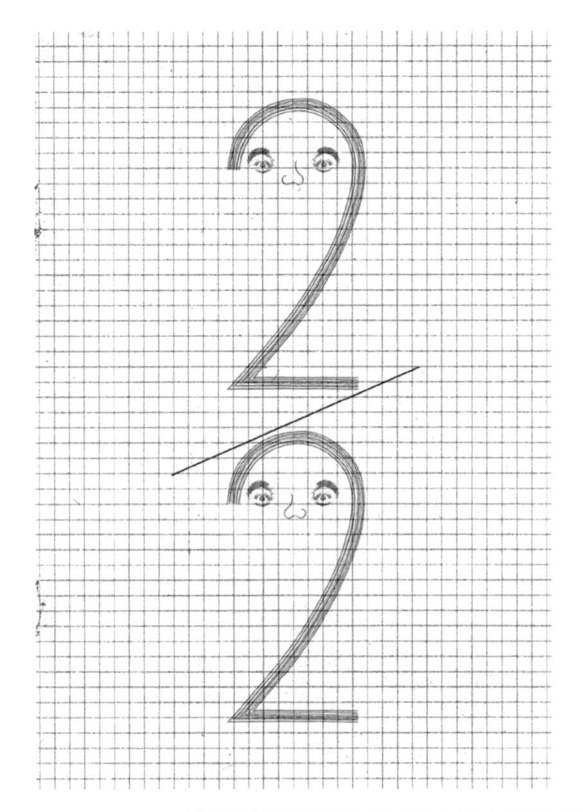

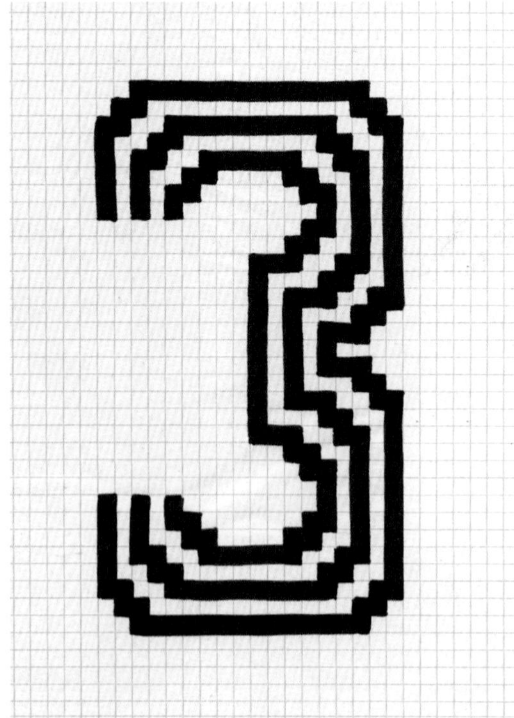

01

设计者：Leonardo Sonnoli，Sara Fanelli，George Hardie

作品名：*If You Could Collaborate*

计数设计上的合作

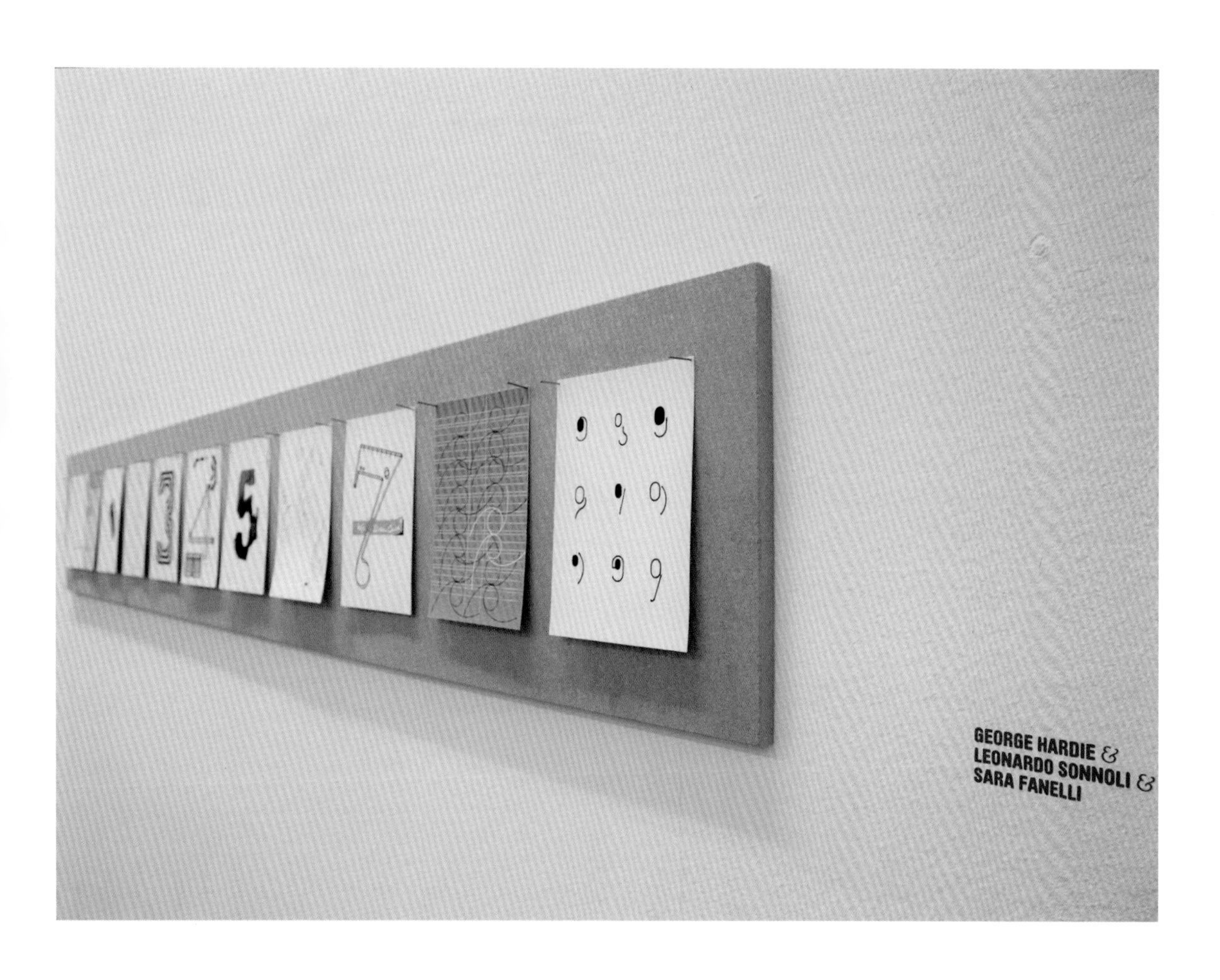
GEORGE HARDIE &
LEONARDO SONNOLI &
SARA FANELLI

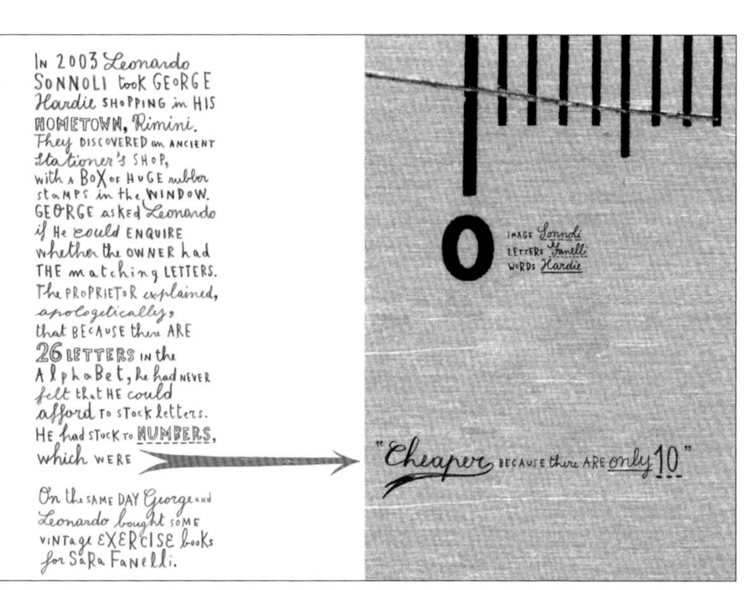
In 2003 Leonardo Sonnoli took George Hardie shopping in his hometown, Rimini.
They discovered an ancient stationer's shop, with a box of huge rubber stamps in the window.
George asked Leonardo if he could enquire whether the owner had the matching letters.
The proprietor explained, apologetically, that because there are 26 letters in the alphabet, he had never felt that he could afford to stock letters.
He had stuck to numbers, which were
On the same day George and Leonardo bought some vintage exercise books for Sara Fanelli.
0
IMAGE Sonnoli
LETTERS Fanelli
WORDS Hardie
"Cheaper because there are only 10"

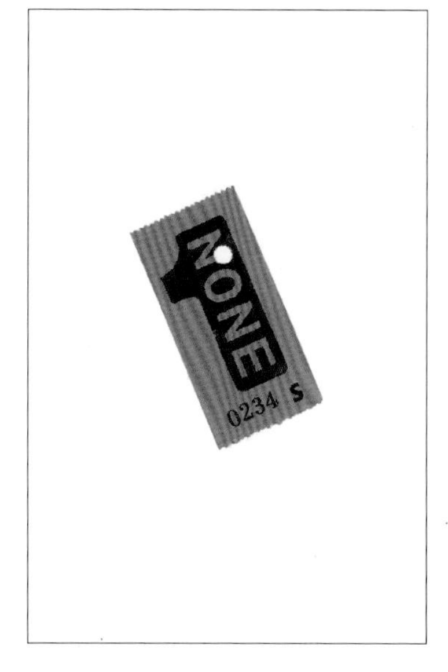
1 NONE
0234 S

05.2.1
遗失与发现

来自莱昂纳多·索诺里合作系列《如果你能合作》的更多实例。

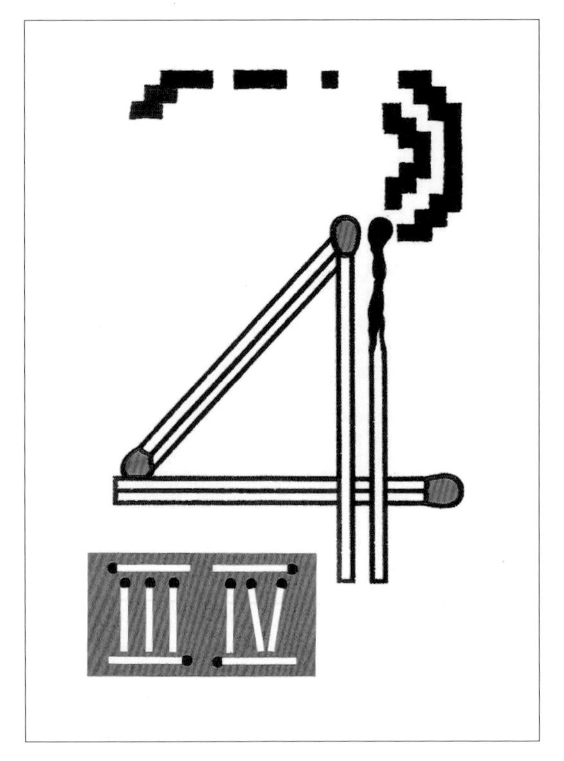

02
设计者：Leonardo Sonnoli, Sara Fanelli, George Hardie
作品名：*If You Could Collaborate*
计数设计上的合作

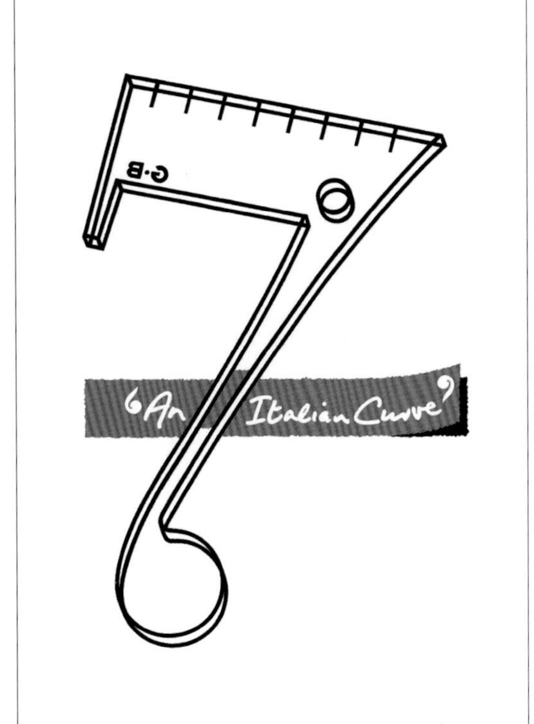
'An Italian Curve'

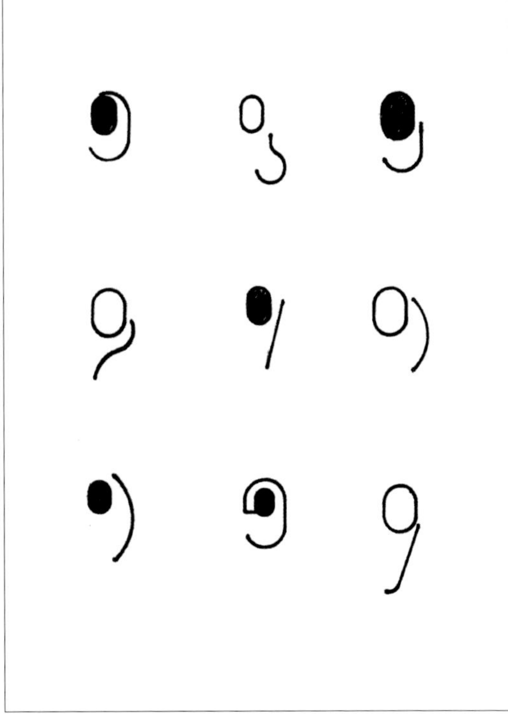

05.2.2
图形元素

《菲尔·哈里森奖》是一款一次性的定制展示件，受索尼电脑娱乐公司欧洲公司（SCEE）委托设计，在索尼电脑娱乐公司全球总裁离职时作为礼品赠送。该设计庆祝他在索尼游戏机平台开发中作为关键人物经历了第15个年头。奖品采用丝网印刷印制于六层有机玻璃之上，装在一个亚克力箱形框架内，通过结合游戏机平台进化的图形元素来形成数字“15”。

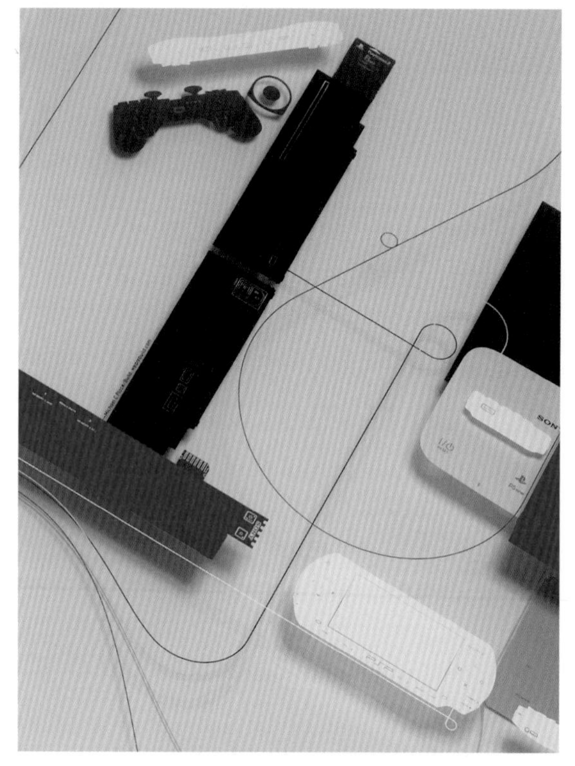

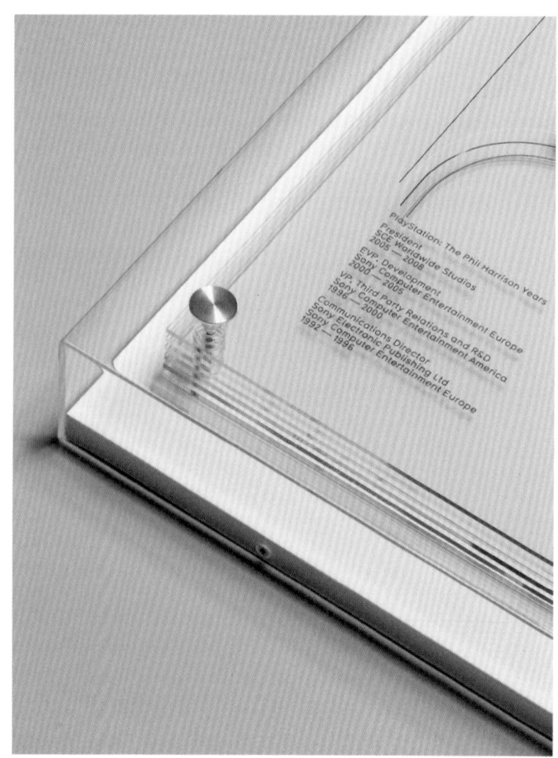

▶ 01

设计者：Build Design
作品名：*Phil Harrison Award*
受索尼电脑娱乐公司欧洲公司（SCEE）委任设计

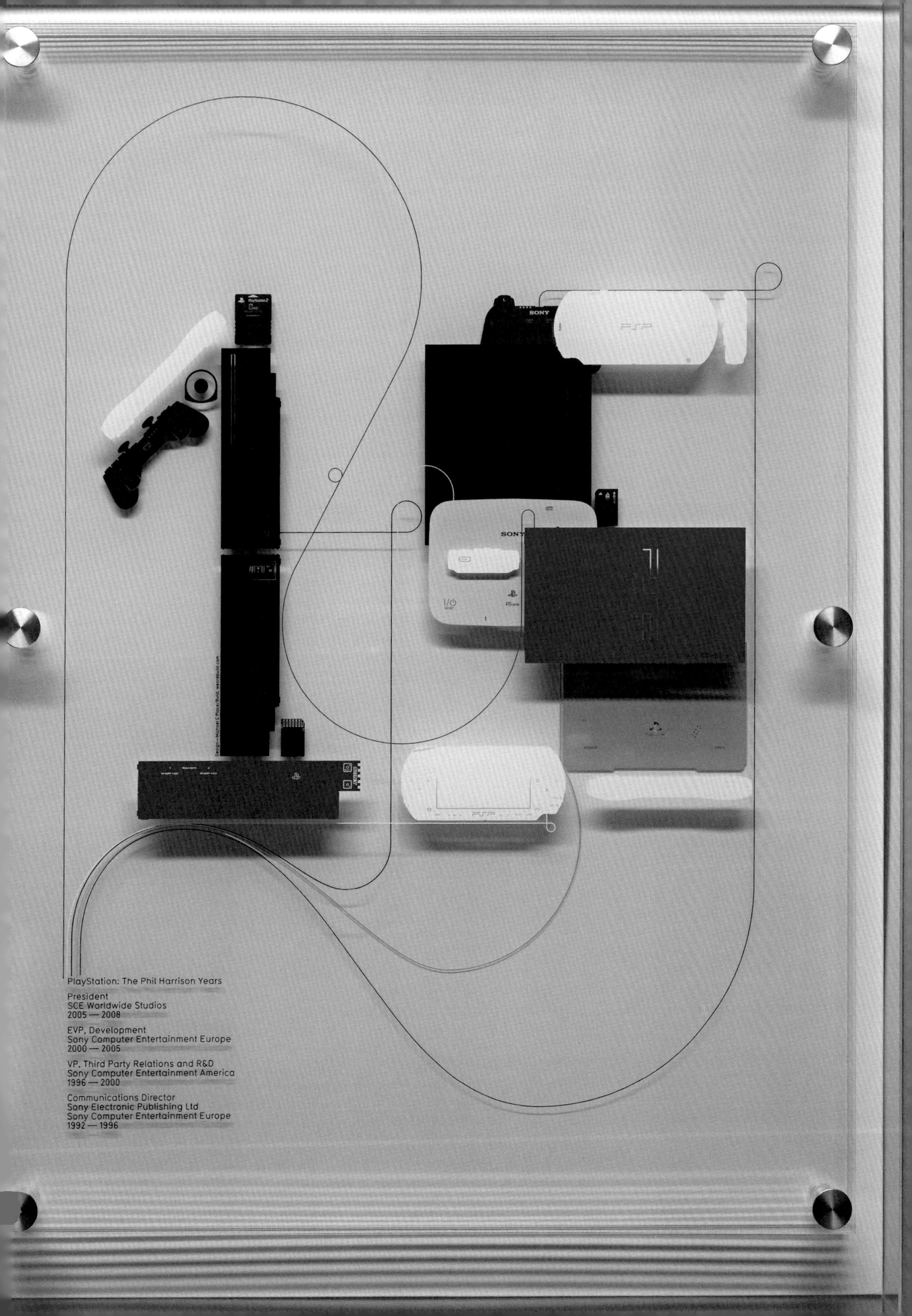
PlayStation: The Phil Harrison Years
President
SCE Worldwide Studios
2005 — 2008
EVP, Development
Sony Computer Entertainment Europe
2000 — 2005
VP, Third Party Relations and R&D
Sony Computer Entertainment America
1996 — 2000
Communications Director
Sony Electronic Publishing Ltd
Sony Computer Entertainment Europe
1992 — 1996
PSP
SONY

05.2.3
空间里的数字

“SEA”设计公司与现代办公家具制造商阿伦德（Ahrend）并肩作战，为《4–2》加工范围内的工作台、会见桌和会议桌制作宣传刊物。此种艺术主导的图像不仅突出了家具系统的灵活性，也巧妙地配置成数字“4”和数字“2”的形式。

01
设计者：SEA Design
作品名：*Ahrend Four-Two*
宣传材料

Four_Two
ahrend
Bench Seat
Solo Table
1 Table
1 Accessory
1 Worktop
Multi Work Bench
3 Basic Modules
4 Connection Modules
2 Work Returns
Conference Table
2 Modules
1 Connection Module
3 Work Tops
Work Bench
2 Modules
2 Work Panels
4 Work Tops
2 Low Storage

05.2.4
划痕与燃烧

“五角设计”为电影《26——五十年的监禁》设计的海报，将优雅的最小化设计与夸大的红色数字“26”相结合，数字“26”与在监狱里度过50年计数的粗糙划痕相冲击。

煞费苦心地采用火柴棒来构造《连线》杂志上某篇文章的开场图片《15 One Two》，就在它们全部焚毁之前，给人一种从结构性空间到字体的形式感。

01

▸ 01

设计者：Pentagram
作品名：*26*
电影宣传海报

▸ 02

设计者：Doyle Partners
作品名：*15 One Two*
《连线》杂志的插图

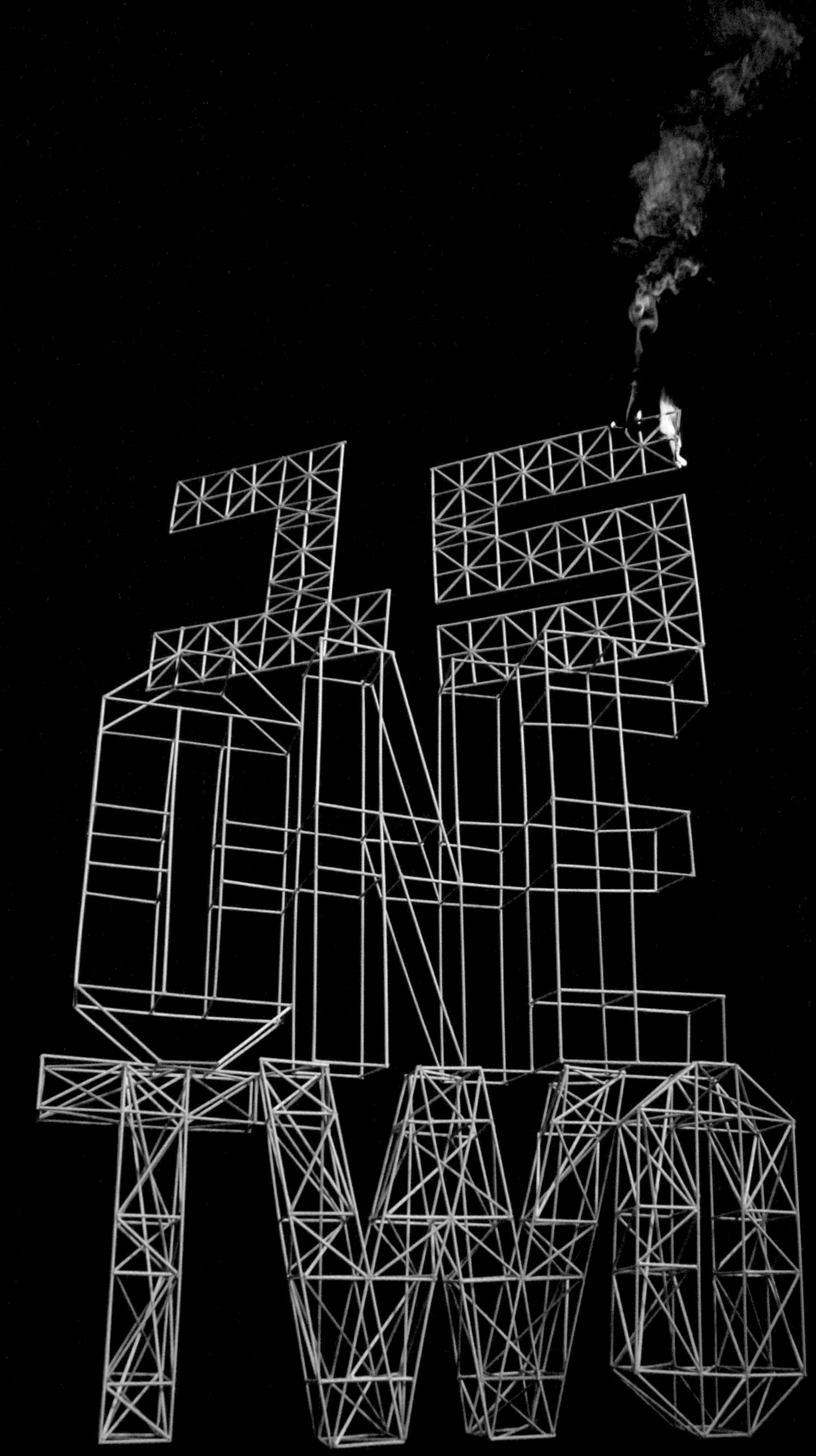

02

05.2.5 数字无处不在

美国建筑与设计杂志《大都会》为庆祝其办刊30周年，特别委托“道尔合伙人”设计的一个封面。封面图片由一大张卡片构成，卡片以这样一种方式折叠形成数字“30”。每期的发行日顺着该结构的整个边缘印制。

《精致的时钟》是2008年由“Brazilian grantholder”在“贝纳通研究中心”开发的一个研究项目。组成《精妙的时钟》的数字取自全世界人民日常生活中每天能看见的，拍摄并上传的事物。时钟建立在一个在线数据库中作为一个Web2.0网站、一个苹果手机应用程序和一系列特定场地的设备而存在。

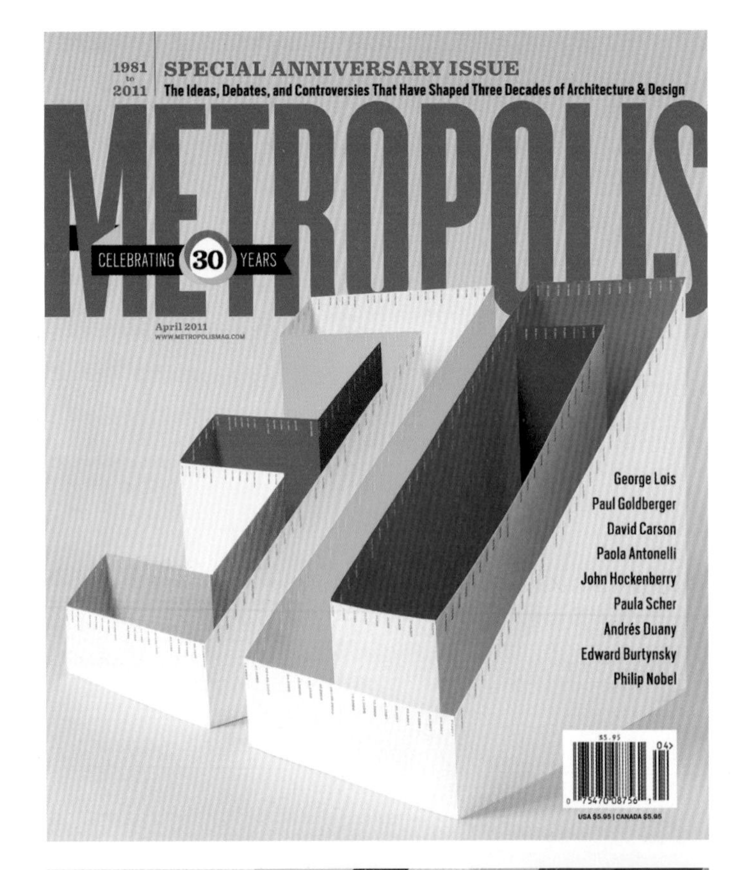

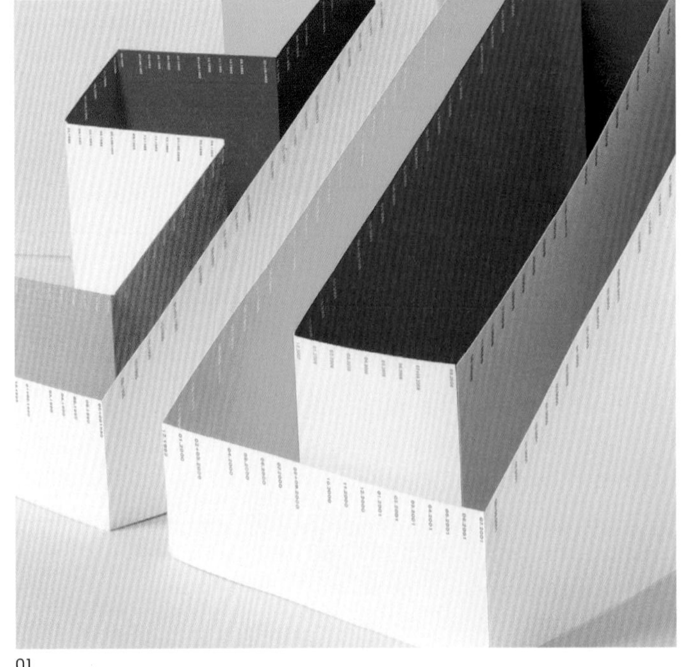

01

▸ 01

设计者：Doyle Partners
作品名：*30 Years Anniversary Cover*
《大都会》杂志

▸ 02

设计者：Fabrica / Joao Wilbert
作品名：*Exquisite Clock*
一款交互式装置，基于网站和应用程序的时钟，该装置艺术见于伦敦维多利亚与艾伯特博物馆（V&A Museum）的《解码：数字设计感》

▸ 03

设计者：Fabrica / Joao Wilbert
Oriol Ferrer Mesia
作品名：*Exquisite Clock*
一款交互式装置，基于网站和应用程序的时钟
设备详图

▸ 04

设计者：Fabrica / Joao Wilbert
作品名：*Exquisite Clock*
一款互动装置，基于网站和应用程序的时钟
图片来源于网站

02

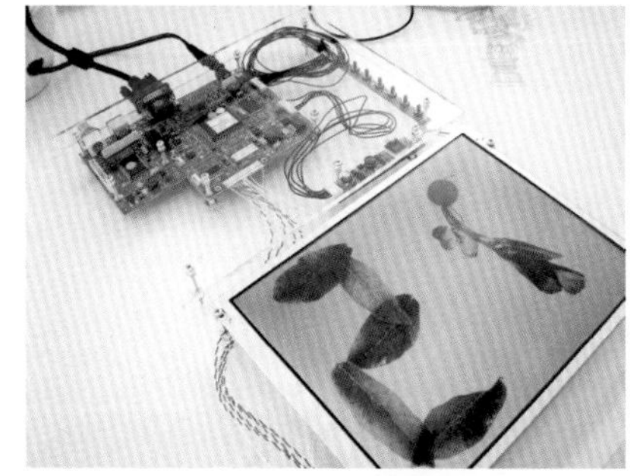

03

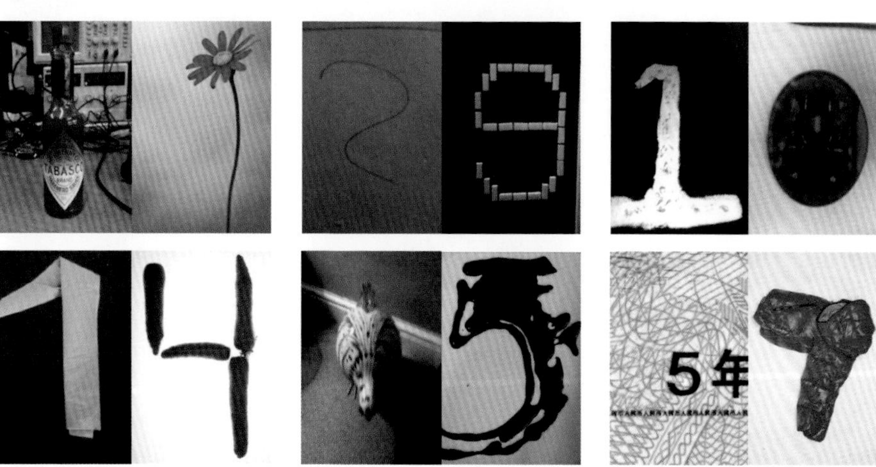

04

05.2.6
现代与古典

“邦奇”设计将纽约电话区码“212”合并至《纽约市1212》的标志，《纽约市1212》是马里奥·德里克（Mario Delic）举办的摄影作品展。针对B1折叠式邀请卡/海报，设计者将标识喷溅于一侧；另一侧则展示德里克的照片。标识的字体具有一种受损与腐蚀的特性，表现出作品展的摄影风格。

印制于深黑色的卡片之上，《欧元塔2》是庆祝欧元塔，即萨格勒布摩天大楼建成两周年的邀请卡。其特点在于白色的箔片压花，数字“2”以塔的特色签名视窗为基础。公司的标志简洁、时尚且呈直线状，而把数字“2”设计为经典的衬线字体。

01

▸ 01

设计者：Bunch Design
作品名：*1212NYC*
B1邀请卡/海报

▸ 02

设计者：Bunch Design
作品名：*Eurotower 2*
《欧元塔2》邀请卡

02

05.2.7
极简主义

艺术家丹·弗莱（Dan Flavin）拍摄了4根日光灯管，为《宠物店男孩——迪斯科4》设计唱片封面，唱片为宠物店男孩翻唱歌曲混音版的收藏集。内套的特点在于拍摄的不同排列，源于灯管的逐步开启和关闭。

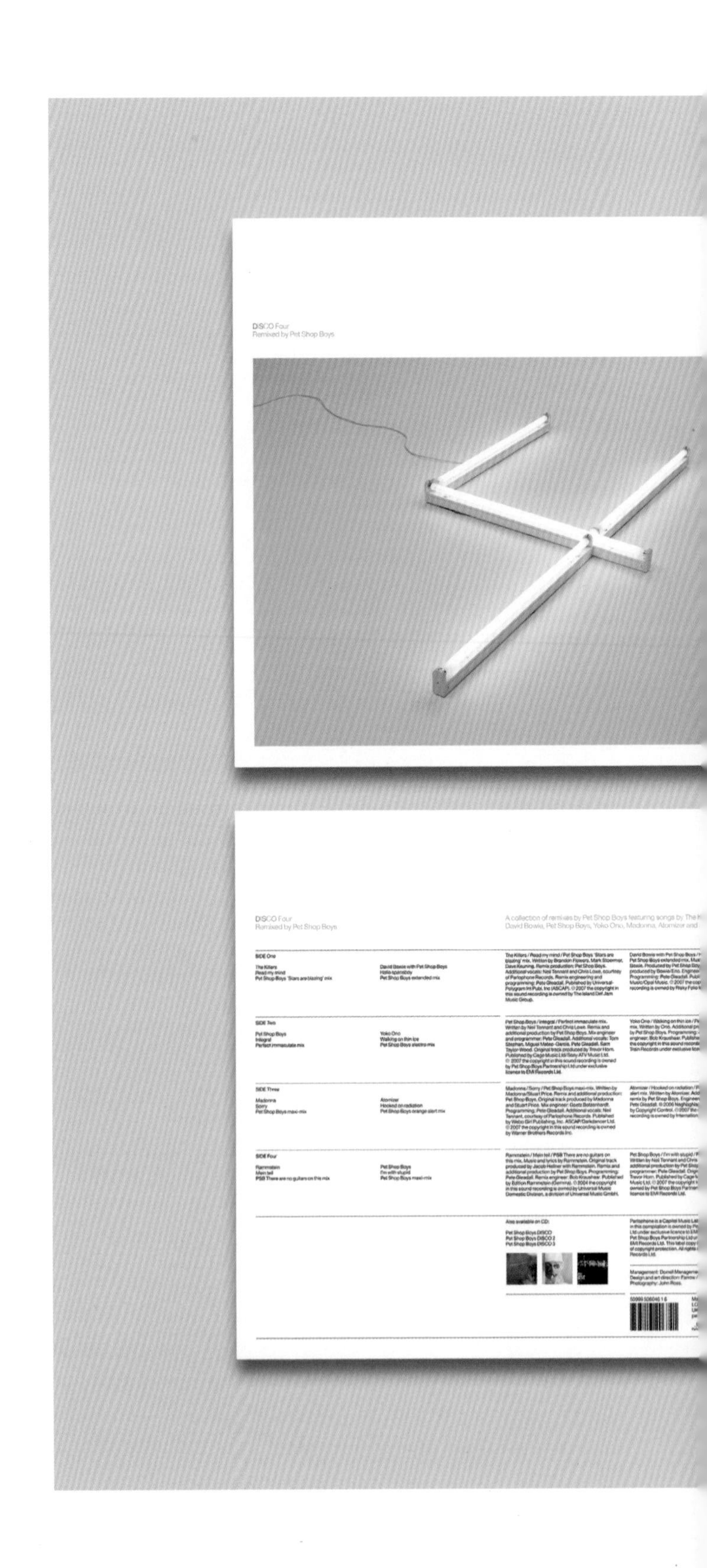

▸ 01

设计者：Farrow Design
摄影：John Ross
作品名：*Pet Shop Boys – Four*
唱片封面及内套

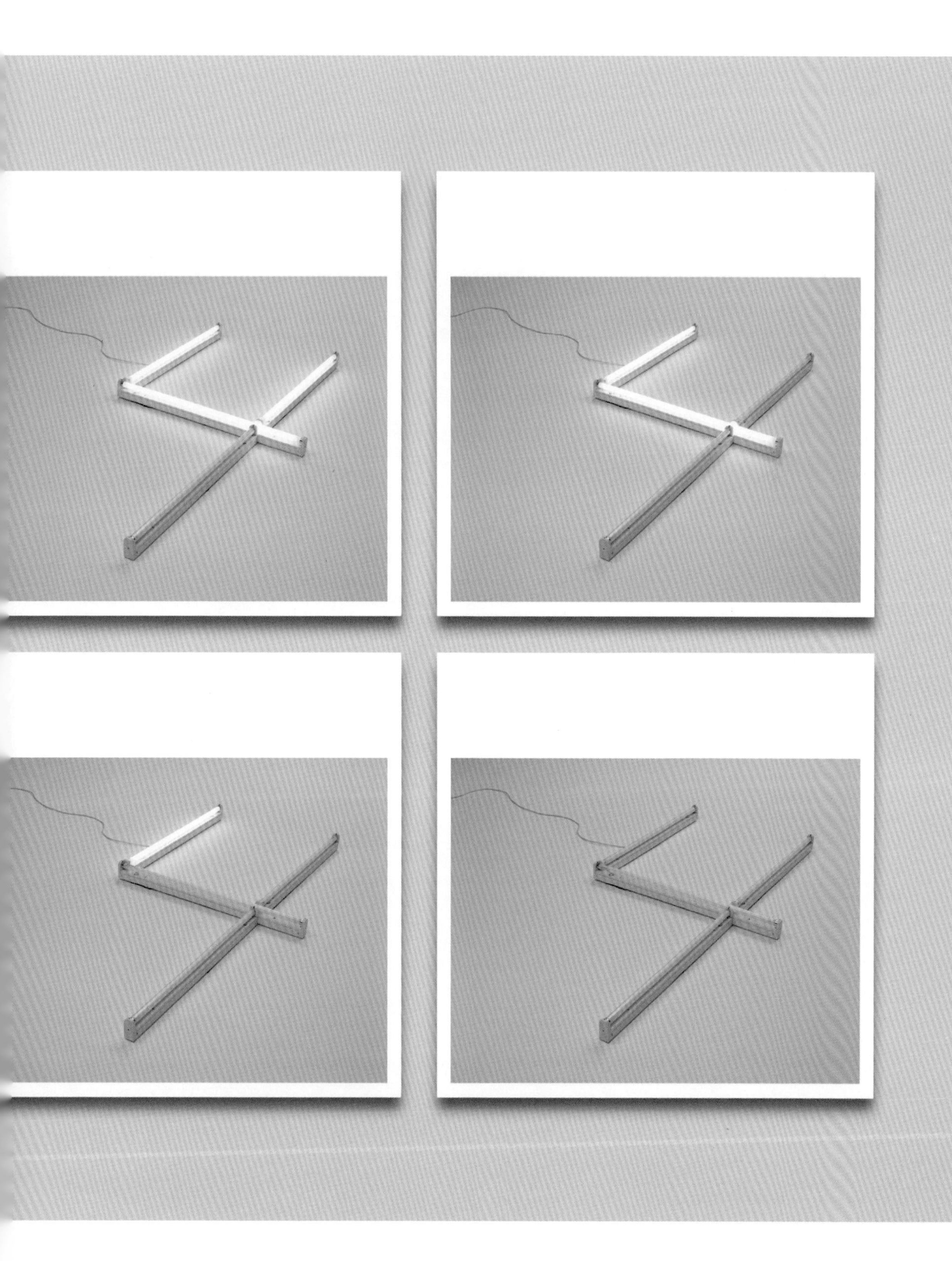

05.2.8
超文本设计

“Form”被邀请为唱片《三首单曲和一张专辑》设计封面，封面需突出一种神秘感。表现出紧张的同时还需表现出英国车库乐队“187 Lockdown”轻松愉快的活力。乐队不希望有他们的照片出现在封面上，因而设计完全围绕形象地表现他们的名字这个鲜明的中心。设计者着重于将数字“187”放置在一系列木板上，与摄影师斯皮洛斯・波利迪斯花了12小时，在伦敦周围他们感觉可以表现出乐队气质的地方进行拍摄。版面和木板是在现场拍摄的，使一系列的封面均具有与众不同的特性。

排版给人以工业的感觉，要素都溶于信息中：“Neue Helvetica”粗体字体印制于专色黄的背景之上，强调了紧张感并传递着“冒险”这一现代主义。

▸ 01
设计者：Form
作品名：*187 Lockdown*
为英国车库乐队：《三首单曲和一张专辑》制作的宣传设计

187
187 LOCKDOWN.
A1: GUNMAN (ORIGINAL MIX)
A2: SOUTHSIDE (ORIGINAL)
A3: IT'S REAL (FEAT. SHOLA PHILLIPS)
B1: YOUNG SON OF KUNG-FU
B2: KUNG-FU (ORIGINAL 187 MIX)
B3: THE WAH
C1: NIGHTMARE ON 187TH STREET
C2: IT'S IN YOUR EYES (FEAT. JANETTE SEWELL)
C3: ALL 'N' ALL (FEAT. D'EMPRESS)
D1: ROOM 1116
D2: SOUTHSIDE (C-KEN REMIX)
D3: THE DON
00:06:19.
00:06:25.
00:06:02.
00:03:20.
00:06:26.
00:06:11.
00:06:18.
00:06:20.
00:05:11.
00:05:44.
00:05:56.
00:06:55.
THE DON
GUNMAN
One way
00:06:03.
00:06:55.
00:05:55.
00:05:45.
00:06:20.
00:03:37.
00:04:49.

05.2.8
超文本设计

《87》是乔纳森·埃勒里（Jonathan Ellery）在伦敦瓦平项目（Wapping Project）的第一次个展上的一部分。作为具有动画和声音的大型装置艺术作品展，把数字置于文本之外，使得数字看起来很抽象。附随的是2 000 200本手工编号版的图书，具有限量版的书套。序列中的每一个数字都具有不同的字体特征，包括最后在20世纪70年代的拉突雷赛印字传输系统目录（Letraset catalogues）中发现的某些奇怪和不拘一格的字体。

02

▸ 02

设计者：Jonathan Ellery
作品名：*87*
瓦平项目（Wapping project）中的装置艺术作品

▸ 03

设计者：Jonathan Ellery
作品名：*87*
展品目录

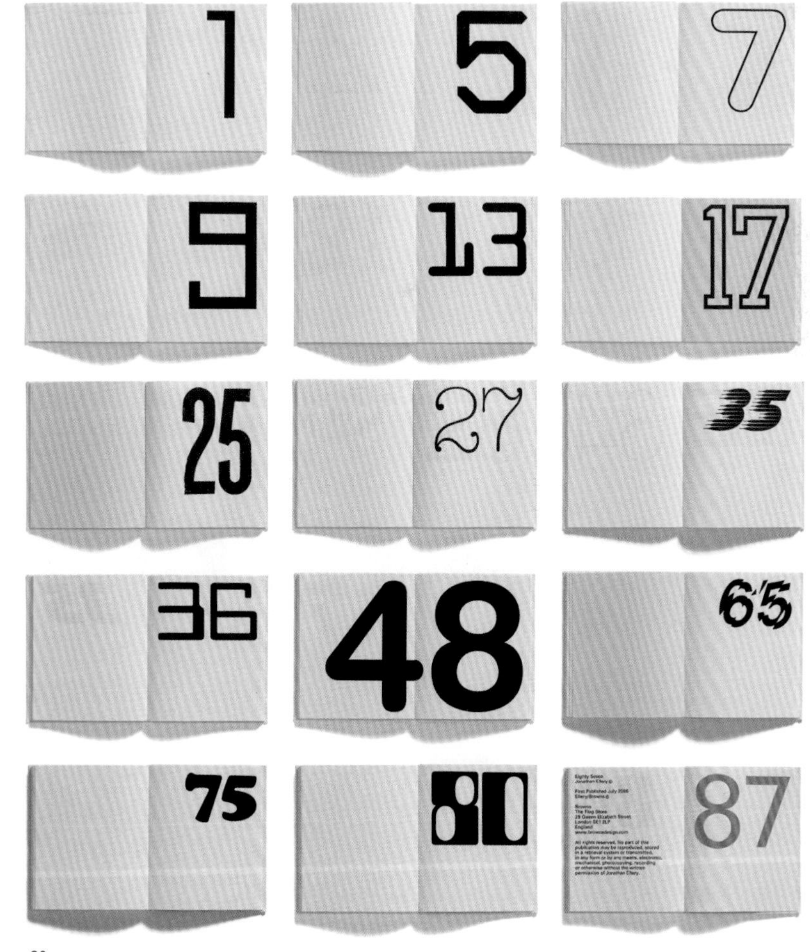

03

05.3
字体设计

05.3.1
组合形式

“Grandpeople”对斯坎迪那维亚时尚杂志《SVA》的重新设计包括开发出的一种新型、独特的标题字体“Framtida”，该字体用于编辑标题和内容列表。将其他字体的碎片粘贴在一起来构造文字。大多数文字的一些变化使得打印页面更具多样性。

“五角设计”为曼哈顿上西城的一处豪华住宅大厦——《88莫宁赛德》设计了身份标识及宣传材料。数字“8”的同心圆设计赋予数字完全的抽象感。

11
222
34
55
67
890

01

▸ 01

设计者：Grandpeople
作品名：*Framtida Typeface*
为《SVA》杂志进行的字体设计

▸ 02

设计者：Pentagram
作品名：*88 Morningside*
宣传手册

02

05.3.2 抽象体

“Stapelberg &Fritz”平面设计工作室设计创作了多种具有实验性和挑战性的字体，重点关注数字特性。

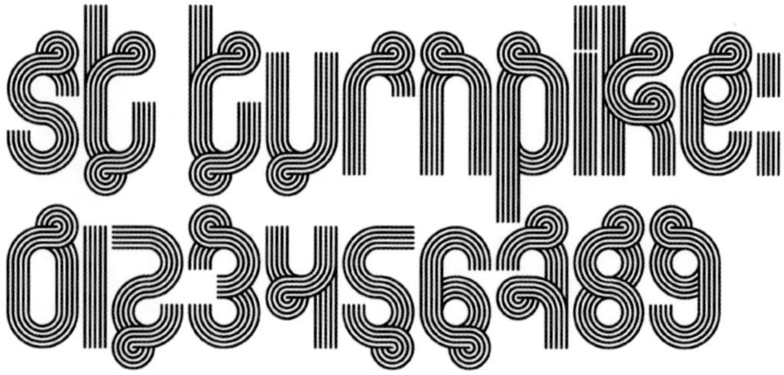

01

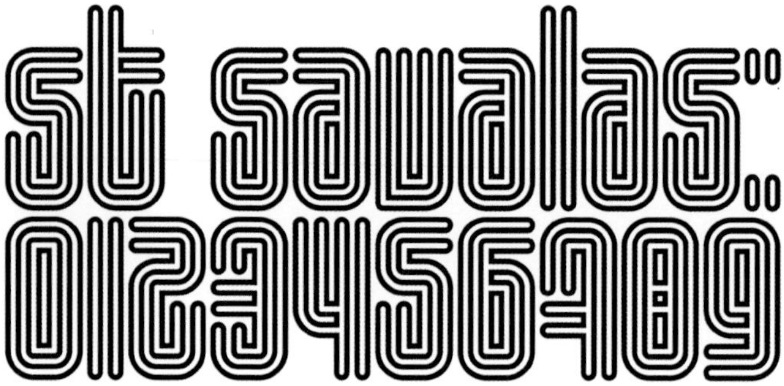

02

03

04

01
设计者：Stapelberg & Fritz
作品名：*ST Turnpike*
由该公司设计的字体

02
设计者：Stapelberg & Fritz
作品名：*ST Savalas*
由该公司设计的字体

03
设计者：Stapelberg & Fritz
作品名：*ST Snail*
由该公司设计的字体

04
设计者：Stapelberg & Fritz
作品名：*ST Deil*
由该公司设计的字体

05
设计者：Stapelberg & Fritz
作品名：*ST Bend*
由该公司设计的字体

06
设计者：Stapelberg & Fritz
作品名：*ST Oplik*
由该公司设计的字体

07
设计者：Stapelberg & Fritz
作品名：*ST Benner*
由该公司设计的字体

sr bend:
0123456789

05

sr split:
0123456789

06

ot bonner:
0123456789

07

05.3.3
源于字母的数字

艾格斯（Eggers）与迪亚博（Diaper）为柏林犹太博物馆称为《10 + 5 = Gott（10 + 5 = God）》的一个展览项目进行设计。

设计者们被希伯来语所吸引，因为希伯来语中没有数字。字母表中的第一个字母“aleph”被用来代表数字“1”，第二个字母“bet”被用来代表数字“2”……这意味着希伯来词语可以同时被视为一连串数字。

在希伯来语中，“10（yod）+ 5（heh）”为“15”，但同时它也表示“yod-heh”这一个词。该词是众多希伯来词汇中表示“神”这一词语中的一个，因而成为展览的名称。为表现希伯来语的这种二元性，设计者们创作出由数字或数字组合形成字母的一种展示字体。

L'acte même de lire – Italo Calvino

dans ville semblable, paru en 1974, Italo Calvino énumère, à la manière de fiches, les nombreuses villes ayant existé d'un bout à l'autre d'un empire (un empire ayant pris de telles dimensions que l'empereur lui-même n'en avait pas l'idée). Dans ce roman, les descriptions spécifiques des villes se suivent simplement, les unes après les autres. Sa forme est celle narrateur, Marco Polo, face à un auditeur, l'empereur, l'empereur qui se contente de l'exhorter à poursuivre son récit. Cette narration continue, donnée comme roman, transforme le lecteur en l'auditeur que fut l'empereur. Dans l'ensemble, aucune des histoires présentées au lecteur ne met en relation les villes entre elles et ne dit quoi ce soit de la structure générale de l'empire. Il est certain qu'à la lecture de ce livre, on ne pourra jamais dire « qu'un jour, un empereur a édifié un tel empire que... » ou que « chaque ville a sa propre forme et sa propre histoire ». À mesure que le lecteur s'enfonce dans le roman, la conscience lui vient du manque d'une ville qui s'est perdue dans sa mémoire ; ou bien, il s'égarera carrément dans la vastitude du texte. D'un côté comme de l'autre, sa conscience en tant que lecteur ne se fera pas à partir du contenu du roman, mais à travers l'acte même de lire.

▸ 01
设计者：Eggers+Disper
作品名：*10+5=Gott*
为柏林犹太博物馆制作的字体设计和展览目录

abcdefghijklmn

6 6 5 9 3 4 5 4 1 1 1 + 7 1 3 2

opqrstuvwxyz

0 6 0 + 7 5 5 1 2 7 3 7 + 7 4 2

the quick brown fox

1 4 3 0 + 7 2 1 5 1 + 7 6 5 0 3 2 4 0 7 + 7

jumps over the lazy dog

1 2 3 6 5 0 7 3 5 1 4 3 1 6 2 4 9 0 5

1234567890

06. 形式

06.1 图形

06.1.1 互连式 204
06.1.2 前期索引 206
06.1.3 字形测试 208
06.1.4 数字形式 210
06.1.5 交错式 212
06.1.6 切割与裁剪 216
06.1.7 排版限制 218
06.1.8 裁片片段 220

06.2 数字

06.2.1 45度 222
06.2.2 数字历史 224
06.2.3 顺序和信息 226
06.2.4 数字主角 228
06.2.5 华丽的排版 234
06.2.6 不成比例 238
06.2.7 数字图像 240
06.2.8 排序 244

06.3 表达

06.3.1 线性 246
06.3.2 焦点 248
06.3.3 书法 250
06.3.4 数字形成的形象 252
06.3.5 色彩游戏 254

06.1
图形

06.1.1
互连式

"Artiva"设计工作室为《莱泰拉22》设计出一款大幅面的折叠期刊和皮夹，《莱泰拉22》是为庆祝设计师阿德里亚诺·奥利维蒂（Adriano Olivetti）的生平而设计的出版物。相互交错的数字"2"，其灵感来自于原版标志性打字机《莱泰拉22》所使用的黑色墨带。

01
设计者：Artiva Design
作品名：*Lettera22*
折叠期刊、海报及DVD

Lettera22
Lettera22
un documentario
di Emanuele Piccardo
su Adriano Olivetti
Cinema Farnese - Persol
Piazza Campo De' Fiori 56
Roma
Lettera22
Lettera22
un documentario di Emanuele Piccardo su Adriano Olivetti
Lettera22

06.1.2
前期索引

《菲利普·梅茨（Philip Metz）印刷品残卷》是菲利普·梅茨收藏品的艺术目录，包括一本法式折叠护封。

01

设计者：Stapelberg & Fritz
作品名：*Sammlung Philip Metz*
艺术品目录

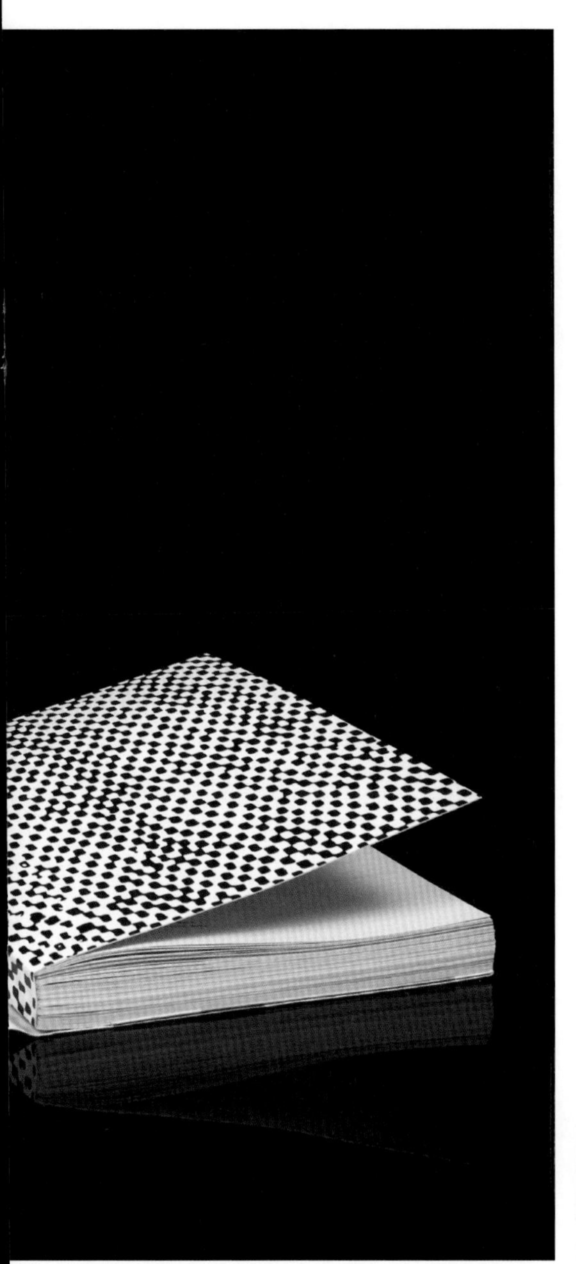

X
A.01
09
/B.01
005
C.01
11
Z

B.01
I WISH
I WAS
2007/
09

06.1.3
字形测试

为颂扬七名伟大影像制作人的作品，“SEA”设计工作室为《自然的》——一系列来自塔里斯·拉塞尔/GF.史密斯的作品设计了宣传方案。这里展示的是字体设计师布鲁诺·玛格（Bruno Maag）的作品。

鉴于之前的版本更多的是基于图像，这里突出的是设计师的Aktiv字体。“X7”指的是发行编号“7”。

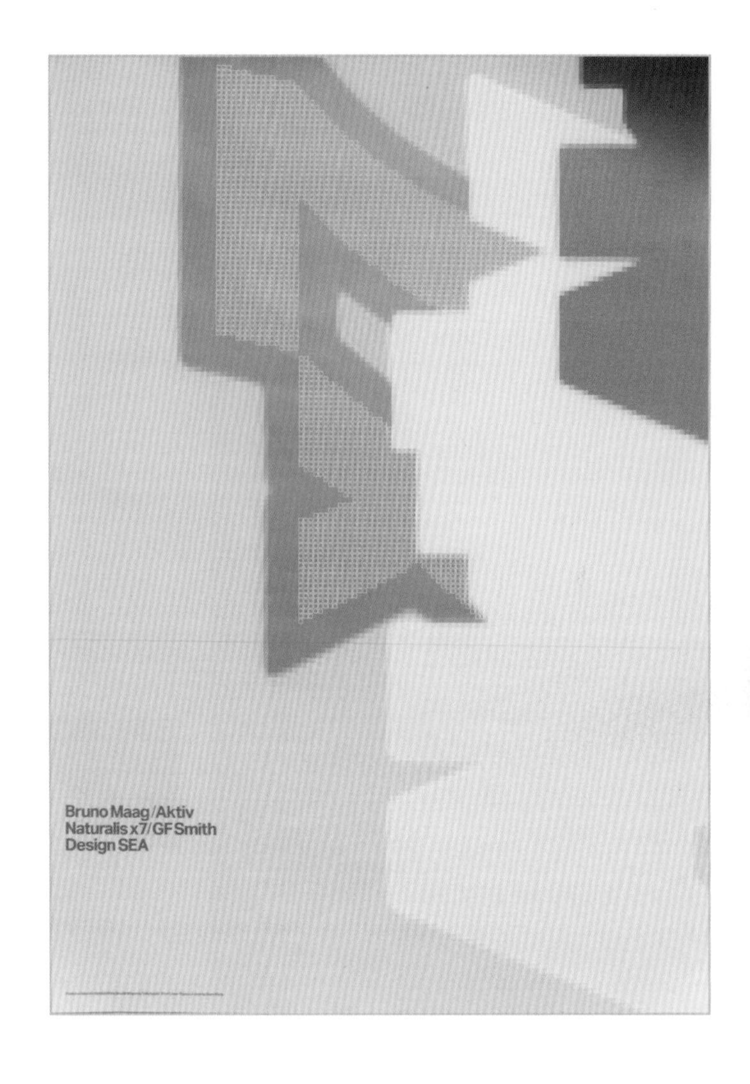

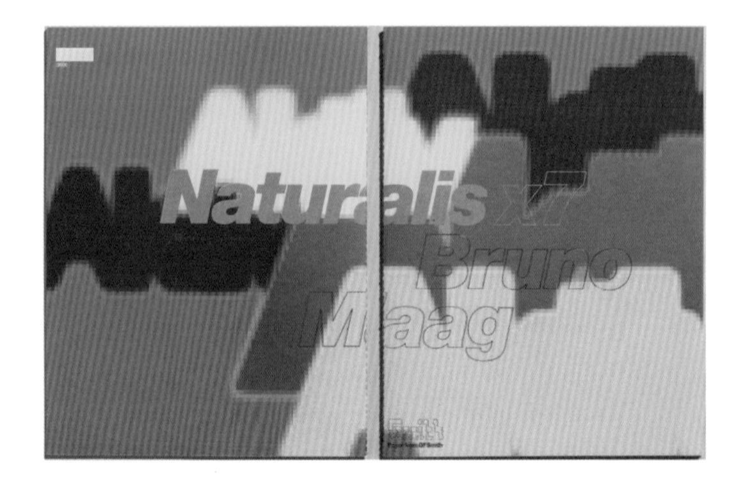

01

设计者：SEA Design
Tullis Russell / GF Smith
作品名：*Naturalis*
宣传活动、海报及宣传册设计

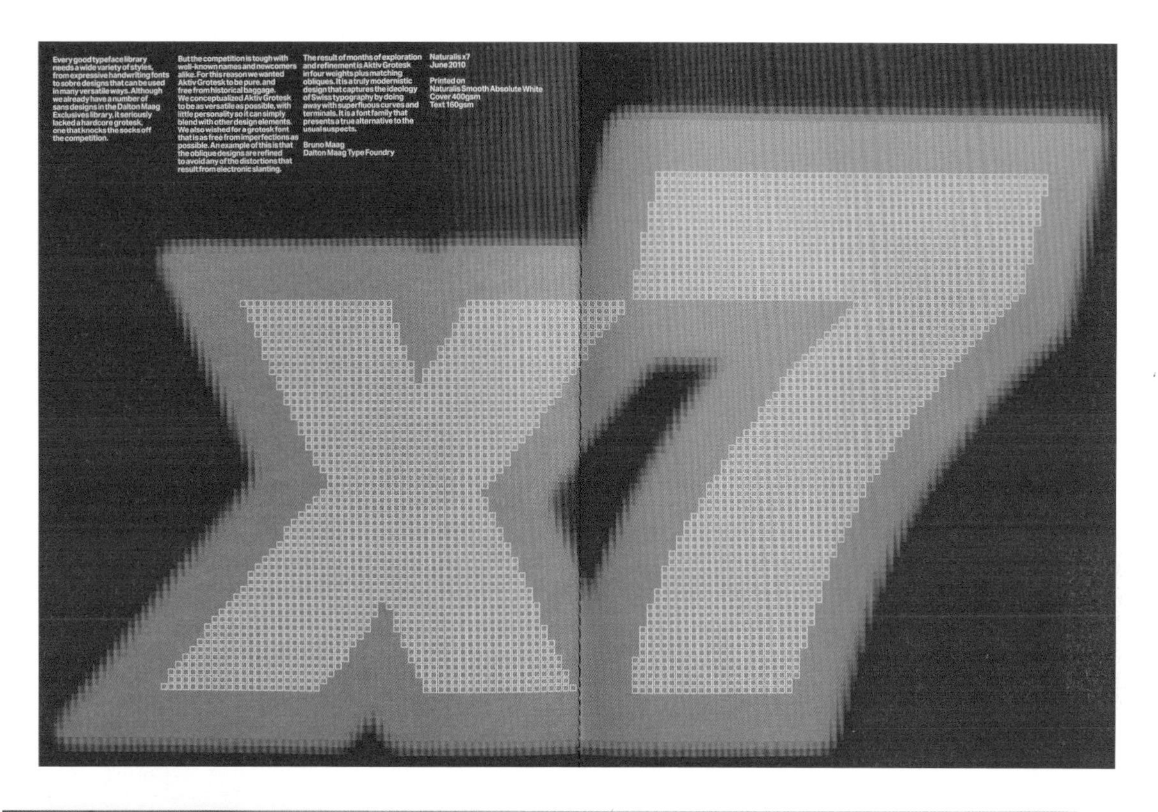
Every good typeface library needs a wide variety of styles, from expressive handwriting fonts to sobre designs that can be used in many versatile ways. Although we already have a number of sans designs in the Dalton Maag Exclusives library, it seriously lacked a hardcore grotesk, one that knocks the socks off the competition.
But the competition is tough with well-known names and newcomers alike. For this reason we wanted Aktiv Grotesk to be pure, and free from historical baggage. We conceptualized Aktiv Grotesk to be as versatile as possible, with little personality so it can simply blend with other design elements. We also wished for a grotesk font that is as free from imperfections as possible. An example of this is that the oblique designs are refined to avoid any of the distortions that result from electronic slanting.
The result of months of exploration and refinement is Aktiv Grotesk in four weights plus matching obliques. It is a truly modernistic design that captures the ideology of Swiss typography by doing away with superfluous curves and terminals. It is a font family that presents a true alternative to the usual suspects.
Bruno Maag
Dalton Maag Type Foundry
Naturalis x7
June 2010
Printed on
Naturalis Smooth Absolute White
Cover 400gsm
Text 160gsm
x7

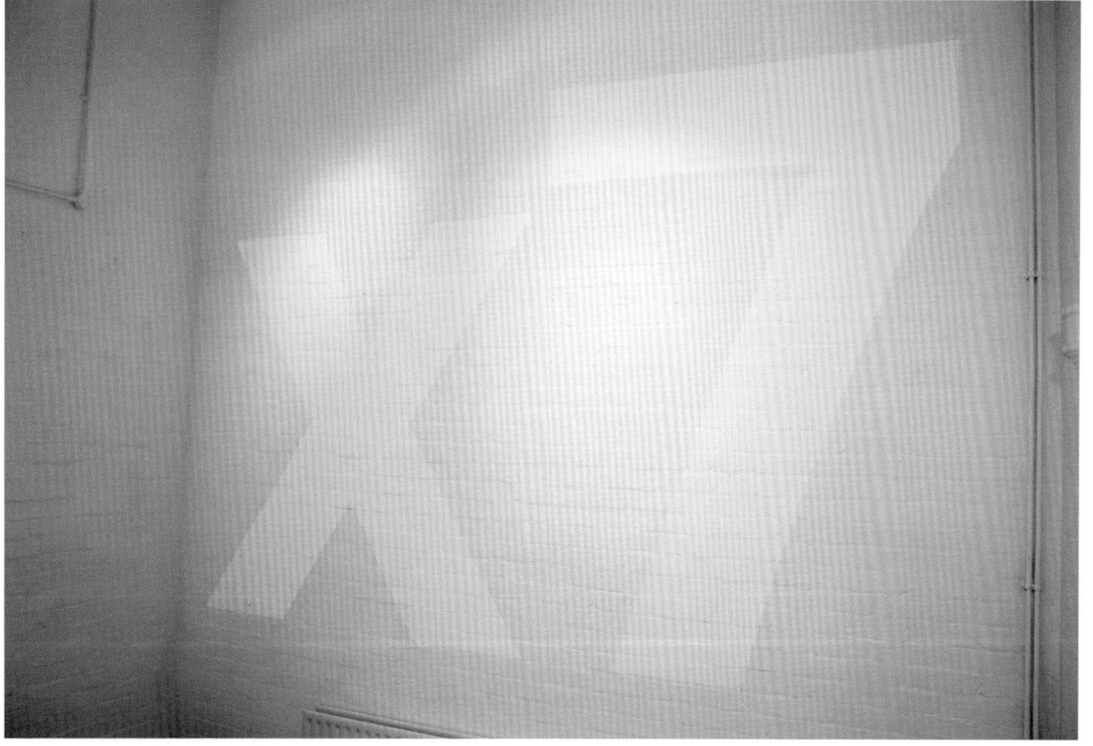

06.1.4 数字形式

马尔科姆·克拉克（Malcom Clarke）为伦敦格雷斯古董中心（Grays antique centre）定制的30周年纪念日邀请卡彰显的是其作为一站式古董商场的地位（其品牌下有200名个体卖家）。在具有200个单元的像素网格的A5纸上获得了这种效果。从众多卖家处拍摄的实物被置于网格中，以创造出数字“30”的一种版式/数字式图形。

“Artiva”设计工作室为“Dopo la rivoluzione”设计出一种宽幅面折叠式期刊、海报以及DVD皮夹。这一图形关注的是排版上的伸展，这里数字以某种裁剪不整的方式表现出文字内的负空间。

2009年举办的《第20届肖蒙国际海报与平面设计节》，组委会邀请来自世界各地的20名设计师每人制作一份海报。莱昂纳多·索诺里（Leonardo Sonnoli）设计了由CMYK字试纸的多个元素组成的巨大数字“20”。他创作了海报的4个版本，每个版本用一种颜色印刷：黑、金属铜、萤光粉及反射蓝。每一版本都呈现出不同的印刷方式。

01

02

▸ 01

设计者：Malcom Clarke
作品名：*30th Anniversary of Grays Antiques*
邀请卡

▸ 02

设计者：Artiva Design
作品名：*Dopo la rivoluzione*
折叠式期刊、海报及DVD

▸ 03

设计者：Leonardo Sonnoli
作品名：*20th edition of the Festival International de l'Affiche et du Graphisme de Chaumont Poster*
海报

20e Festival International de l'Affiche et du Graphisme de Chaumont
du 16 mai au 14 juin 2009

Ville de Chaumont.
Avec le soutien du Conseil général de la Haute-Marne,
du Conseil régional Champagne-Ardenne
et de la Direction Régionale des Affaires Culturelles /
Ministère de la Culture et de la Communication.

leonardo sonnoli

03

06.1.5
交错式

“SEA”设计工作室为伦敦领先的丝印工作室“K2”设计的标识和海报。字母“K”与数字“2”由水平线条组成，相互交错形成不同的色彩纹路和组合，以创造出一种灵活和多样化的标志。

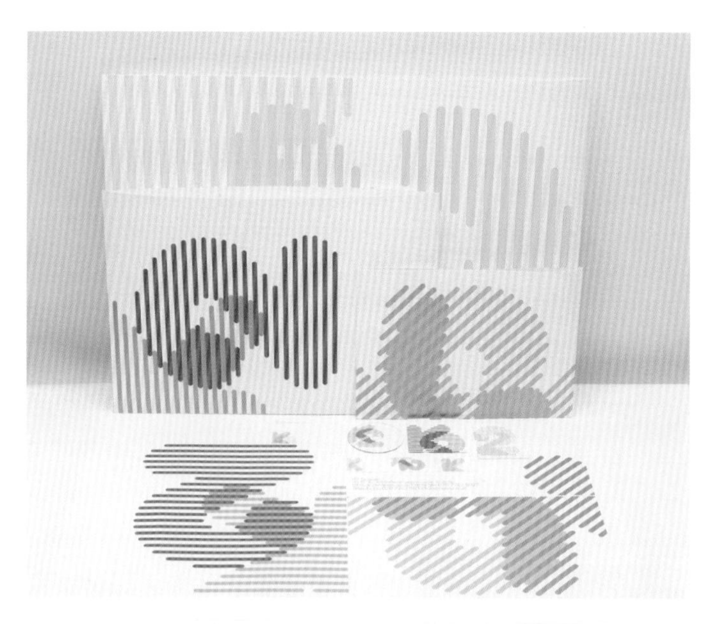

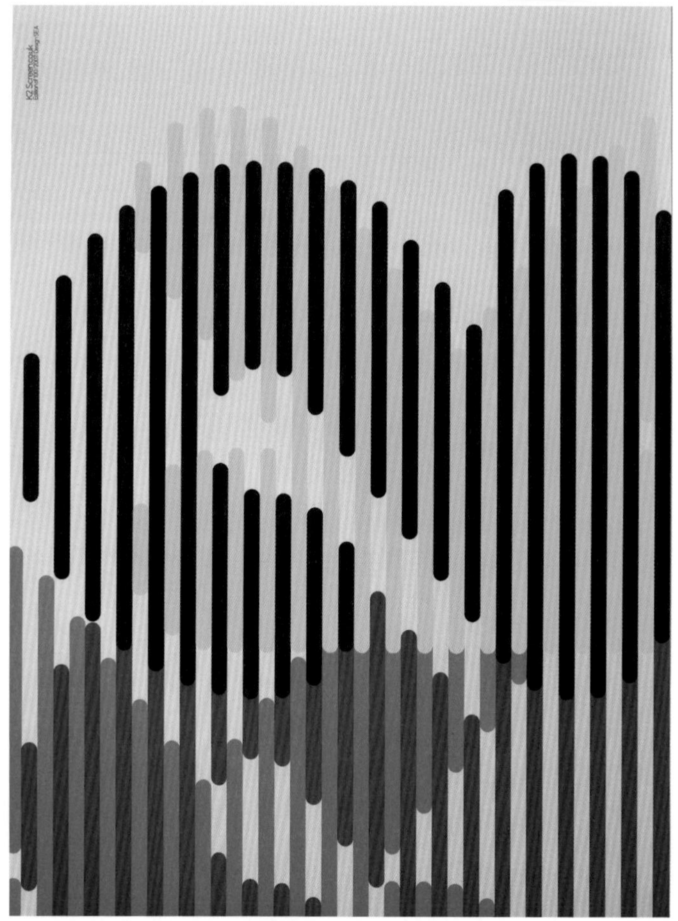

01

设计者：SEA Design
作品名：*K2*
标识与海报

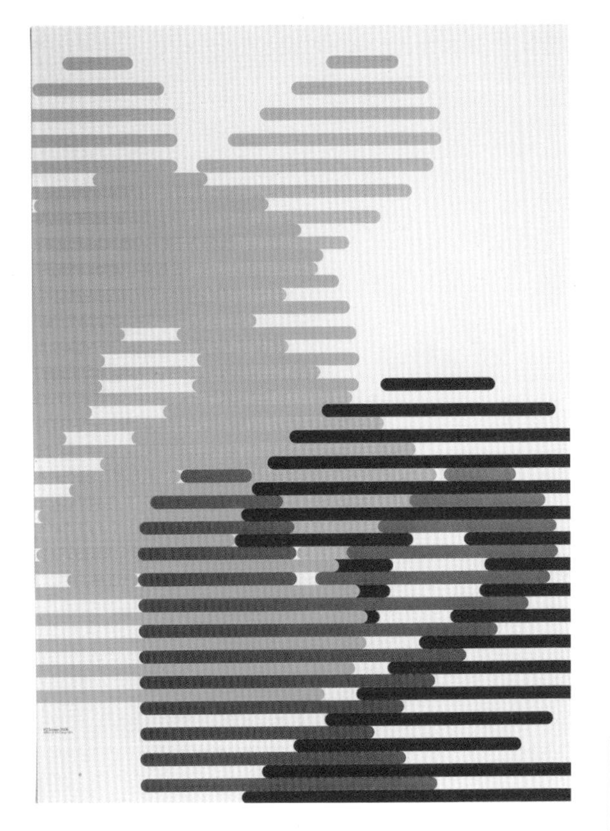

01

06.1.5
交错式

“Remake”设计公司为《志愿律师们为了艺术》设计的年报封面表明该报告已持续两年。大胆地使用粗线条将上一年的数字灵活地缩减使其穿过本年度的数字，之后两数字会叠加。

汤普森·布兰德·帕尔纳（Thompson Brand Parner）为印刷商金斯博里·普雷斯设计的2009年及2011年《新年卡片》结合箔片凸印、压花及传统的平面印刷术，形成该公司印刷技术及各种不同技术的迷你陈列柜。字母形式与背景线路相互交错。

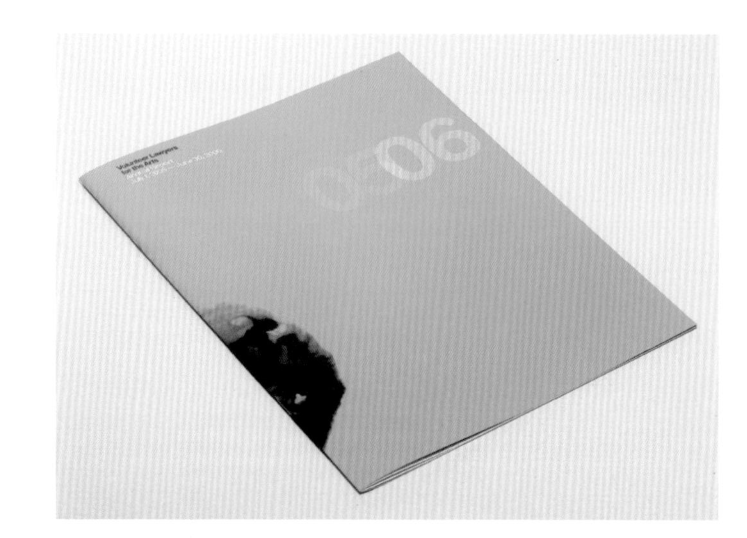

02

▸ 02

设计者：Remake Design
作品名：*Voluntter Lawyers for the Arts*
年报封面

▸ 03
▸ 04

设计者：Thompson Brand Partners
作品名：*Kingsbury Press 2009*
新年卡片

▸ 05

设计者：Thompson Brand Parners
作品名：*Kingsbury Press 2009*
日记簿

▸ 06
▸ 07

设计者：Thompson Brand Partners
作品名：*Kingsbury Press 2011*
新年卡片

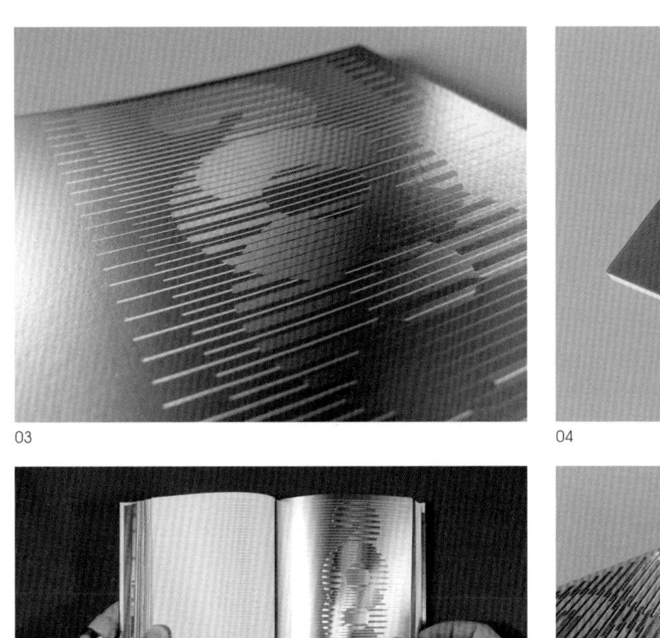

03

04

05

06

07

06.1.6
切割与裁剪

帕劳基金会（Palau Foundation）收藏了有关毕加索的、重要的记录性文件和书目，用于供国内及国际的所有观众参观，同时组织了临时性的展览。该基金会第五年度艺术品目录中的《帕劳身份识别号》采用一种程式化的漏版字体，延长至数字“03”和“08”。该基金会成立于2003年，目录制成于2008年。

《4257》是为日本制药公司“Taiho”设计的公司内部杂志的名称。身份识别号及封面由赫尔穆特·舒密特（Helmut Schmid）设计，商标由特别绘制的连锁数字构成。

01

▸ 01

设计者：Marnich Associates
作品名：*Fundació palau fifth anniversary art catalogue and brochures*
帕劳基金会

▸ 02

设计者：Helment Schmid
作品名：*4527*
“Taiho”制药公司内部杂志

月刊
ヨコヅナ
3月
9

02

palau
03-08
FUNDACIÓ PALAU

Cinc anys de Fundació
2003–2008
palau
03-08
FUNDACIÓ PALAU
CENTRE D'ART
Caldes d'Estrac

06.1.7
排版限制

ISIA（乌尔比诺工艺美术高等学院）是意大利最古老的平面造型设计大学。它坐落在意大利中部美丽的乌尔比诺文艺复兴村。

为推广其课程，莱奥纳多·索罗里设计了5个基于数字的系列海报。采用双面印刷，每一面的特色在于对数字“3”和“2”的不同版式处理：该校学制为学士学位3年，硕士学位2年。用黑色印制于五种不同颜色的纸张之上，海报为悬挂设计，因此数字“3”和“2”并排出现。

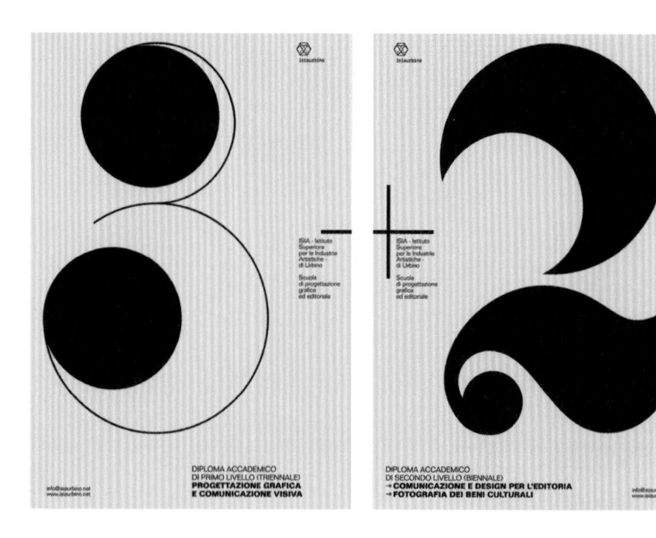

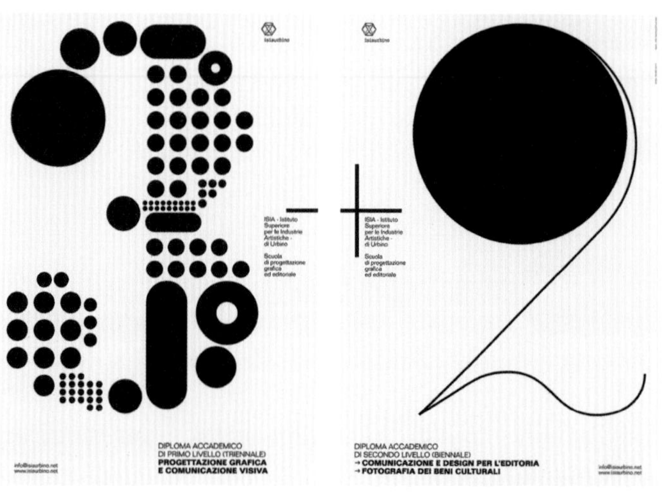

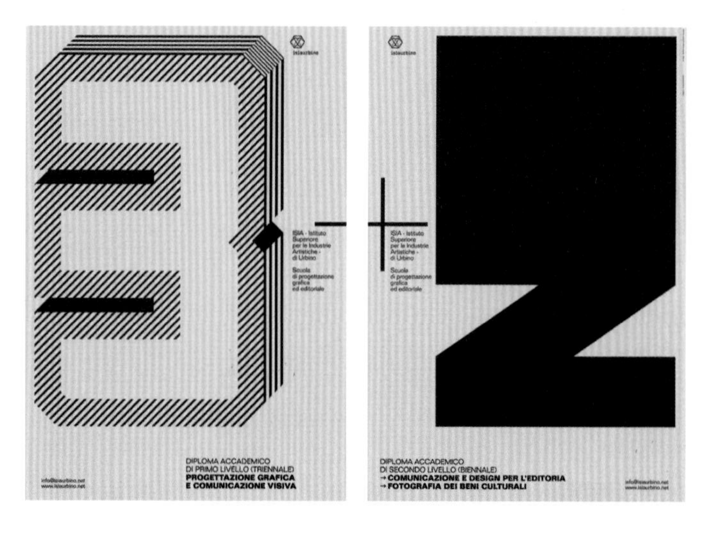

▸ 01

设计者：Leonardo Sonnoli

作品名：*ISIA Posters*

海报系列

isiaurbino
ISIA - Istituto
Superiore
per le Industrie
Artistiche -
di Urbino
Scuola
di progettazione
grafica
ed editoriale
DIPLOMA ACCADEMICO
DI PRIMO LIVELLO (TRIENNALE)
PROGETTAZIONE GRAFICA
E COMUNICAZIONE VISIVA
info@isiaurbino.net
www.isiaurbino.net

isiaurbino
ISIA - Istituto
Superiore
per le Industrie
Artistiche -
di Urbino
Scuola
di progettazione
grafica
ed editoriale
DIPLOMA ACCADEMICO
DI SECONDO LIVELLO (BIENNALE)
→ COMUNICAZIONE E DESIGN PER L'EDITORIA
→ FOTOGRAFIA DEI BENI CULTURALI
info@isiaurbino.net
www.isiaurbino.net

isiaurbino
ISIA - Istituto
Superiore
per le Industrie
Artistiche -
di Urbino
Scuola
di progettazione
grafica
ed editoriale
DIPLOMA ACCADEMICO
DI PRIMO LIVELLO (TRIENNALE)
PROGETTAZIONE GRAFICA
E COMUNICAZIONE VISIVA
info@isiaurbino.net
www.isiaurbino.net

isiaurbino
ISIA - Istituto
Superiore
per le Industrie
Artistiche -
di Urbino
Scuola
di progettazione
grafica
ed editoriale
DIPLOMA ACCADEMICO
DI SECONDO LIVELLO (BIENNALE)
→ COMUNICAZIONE E DESIGN PER L'EDITORIA
→ FOTOGRAFIA DEI BENI CULTURALI
info@isiaurbino.net
www.isiaurbino.net

06.1.8
裁片片段

“Struktur”设计的《单位》包括两本完全相同的线装壁挂式日历，每一本都含有数字“0”至“9”共10张页面，用户翻阅它们便可显示当前日期。每张印制于一种不同颜色的卡片上，裁片以这样一种方式摆放，即数字显示其他数字的下一步片段，一个小型的三角形孔径直穿过前方页面到达后面墙壁。

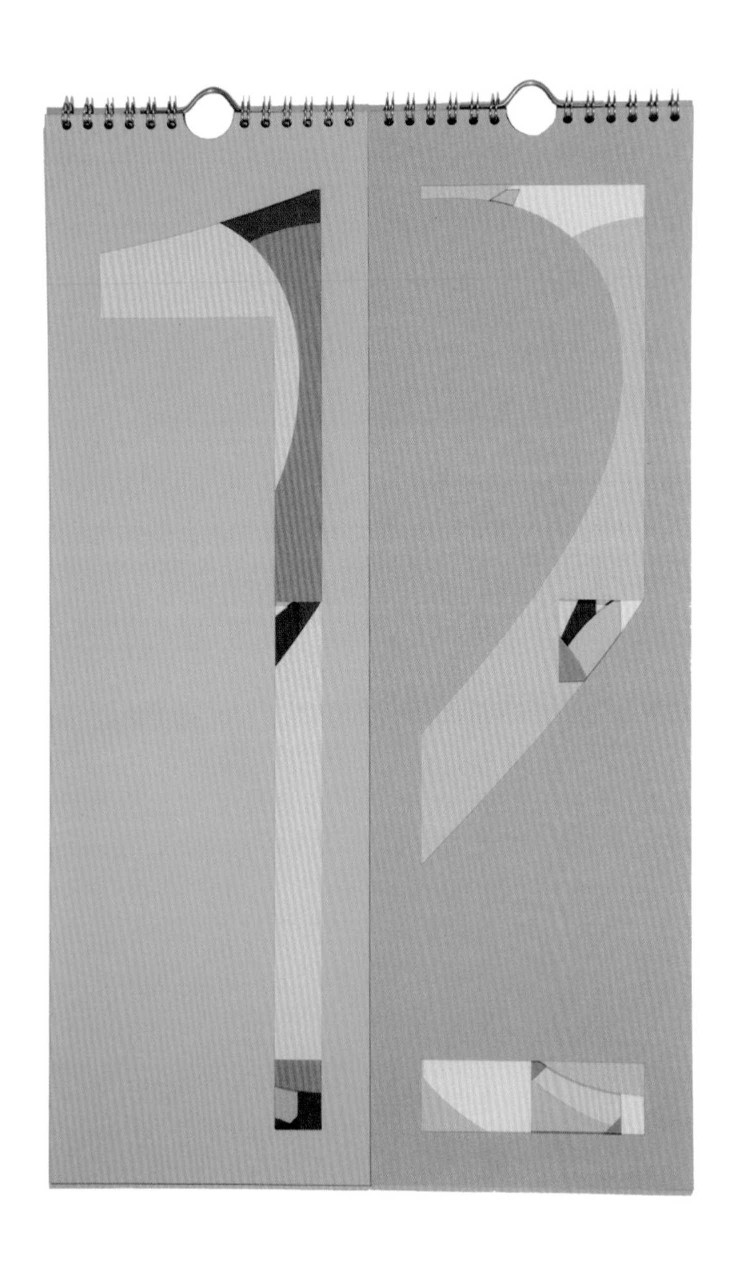

01
设计者：Struktur Design
作品名：*Units*
裁片式日历

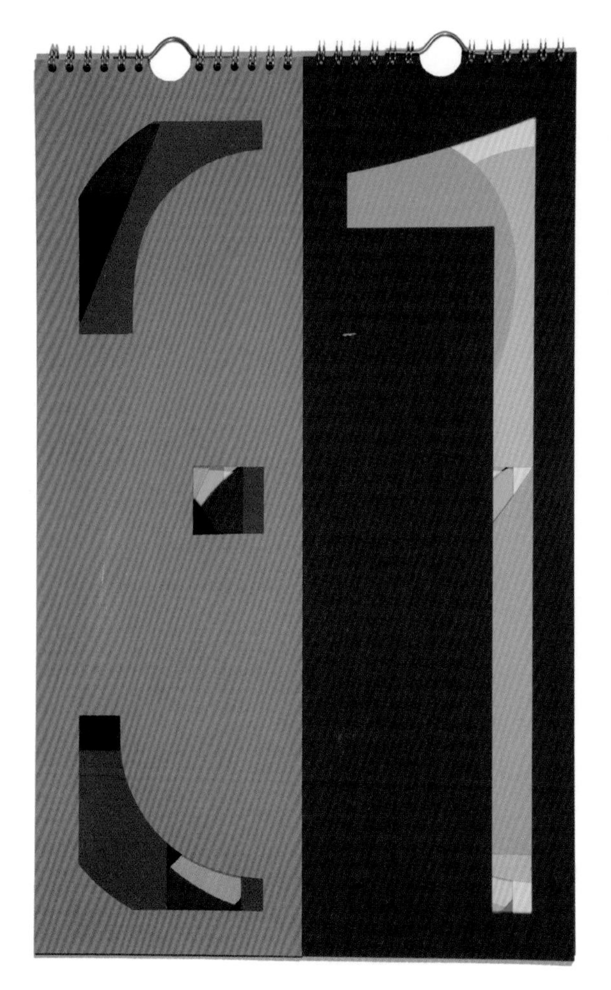

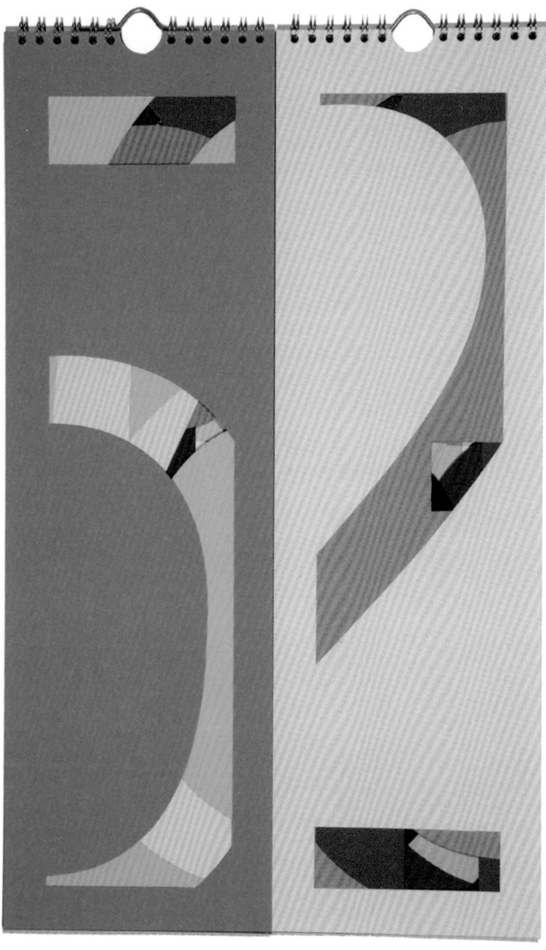

06.2
数字

06.2.1
45度

在零售及住宅地产集团占主导地位的“联盛集团”委托卡特利吉·列文（Cartlidge Levene）为其设计一系列三维图形方案来激励和吸引员工与访客。为公司伦敦新总部的每一楼层设计了一系列不同颜色的框架式环境艺术场景。在这些盒子的表面破口形成基于漏版的数字以创造出一种纵深感和空间感。其他的盒子可以让职员选择各种图像进行填充。

“Artiva”设计公司为瑞典能源公司“45°”设计的形象识别手册将概念简化到其最纯粹的造型——设置成45度角的一根直线。

01

▸ 01

设计者：Cartlidge Levene
作品名：*Lend Lease*
楼层编号方案

▸ 02

设计者：Artiva Design
作品名：*45°*
形象识别手册

Corporate
Identity
Styleguide
Energy Company®
13—6—08
E\Bag®
E\Implant®
E\Pannel®
®

45°
®

02

06.2.2
数字历史

第一批移民被认为是在第9世纪抵达冰岛：最古老的人类居住遗迹发现于雷克雅未克（Reykjavik）和克里苏维克（Krysuvik），根据放射性碳定年法测算距今大约871年。这里的标志及目录着重于这一重要日期以及《雷克雅未克871±2》这一图案：以一种清新和现代的方式看待“人居展览”，与历史性文物形成对比。

▸ 01

设计者：Atelier Atli Hilmarsson
作品名：*ReyKjavik 871±2: The Settlement Exhibition*
标识、会展图案及目录设计

Reykjavík
871±2
Reykjavík
871±2

Reykjavík
871±2
Landnámssýningin
The Settlement
Exhibition

06.2.3
顺序和信息

过去的十年里，“SEA”设计工作室与“GF史密斯”公司一起致力于研究印刷及放于互联网上的所有用于产品宣传的身份标志和印刷品系统。

顾名思义，《主调节器》是一个综合的印刷品系统，它包含装有两本信息调查簿的一个外部蛤壳式匣子和一套4册纸样簿，纸样簿展示了该公司全系列的纸张。每一样本簿前页的大号银色箔片凸印数字使整个包装给人一种秩序感。

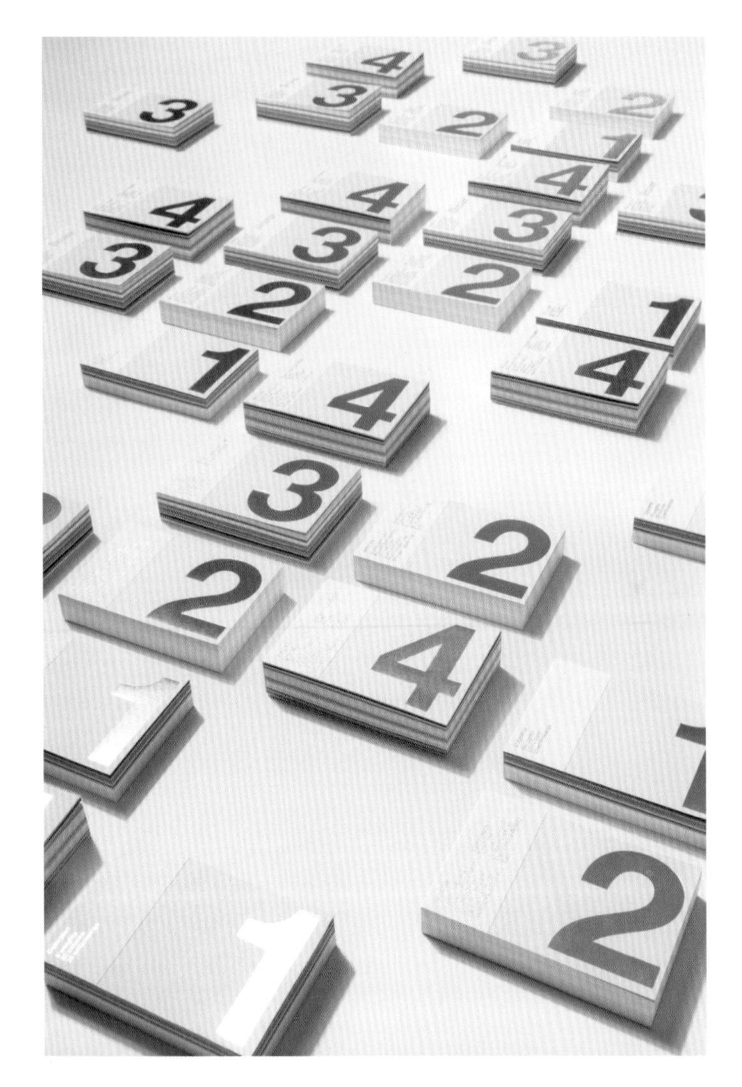

01
设计者：SEA Design
作品名：*Master Selector*
“GF史密斯”纸张挑选信息包

1
2
3
4
Select
Colour
Select
Black
White
1
Colorplan
2
Smooths
3
Textures

Parch Marque
Grammage (gsm)
Sheet Size (mm)
Digital
Colours
90
635x965
460x320
All Colours
176
660x1016
320x460
All Colours
250
660x1016
460x320
All Colours
→4 Specials
Peregrina Classics
Grammage (gsm)
Sheet Size (mm)
Digital
Colours
120
720x1020
320x460*
Metallic-
Silver
Midas
290
700x1000
Anthracite-
Fractal
Tungsten-
Fractal
300
720x1020
320x460*
Midas
305
720x1020
Brushed-
Aluminium
Metallic Gold
Metallic Silver
336
720x1020
Ocean Blue
Ruby Red
→4 Specials
Peregrina Majestic
Grammage (gsm)
Sheet Size (mm)
Digital
120
720x1020
320x460
320x450
210x297
All Colours
250
720x1020
460x320
450x320
210x297
Real Gold
Real Silver
290
720x1020
460x320
450x320
210x297
All Colours
Except
Real Gold
Real Silver
All Embossings All Embossings
→4 Specials

06.2.4
数字主角

2009年是标志着都柏林一家工作室及创作室——“消防站艺术家工作室”成立十周年。“大卫·史密斯”设计工作室将镂空数字图案大胆地应用于庆祝这一事件的图书外封面。

塔拉矿山（Tara Mines）是欧洲最大的铅锌矿山，也是世界上最大的铅锌矿山之一。在过去的30年里，米斯郡纳万郊区（Navan，County Meath）一直在开采并提炼着铅锌矿石。在2008和2009年，艺术家蒂姆·达勒姆（Tim Durham）获得了进入该矿山所有区域的特殊机会，他对地面以及灯光下的地下所有可见情况都进行了拍摄。形成的对工业景观和内部设计的记录造就了“至日艺术中心”的一场展览。附带的图书《650-1575》得名于矿山的深度。大号的粗体数字环绕书本形成醒目的封面。数字下部超出页面以显示地下的概念。

《原料的选择》这一海报号召设计者们加入2003年一项学生设计大赛，海报焦点在于较大数字“03”。

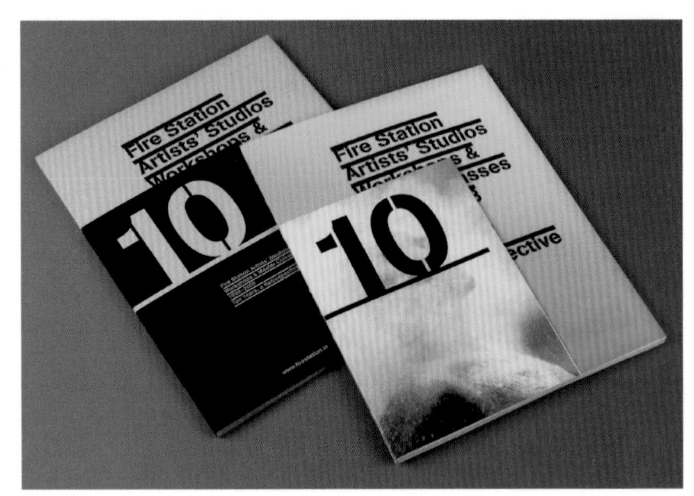

01

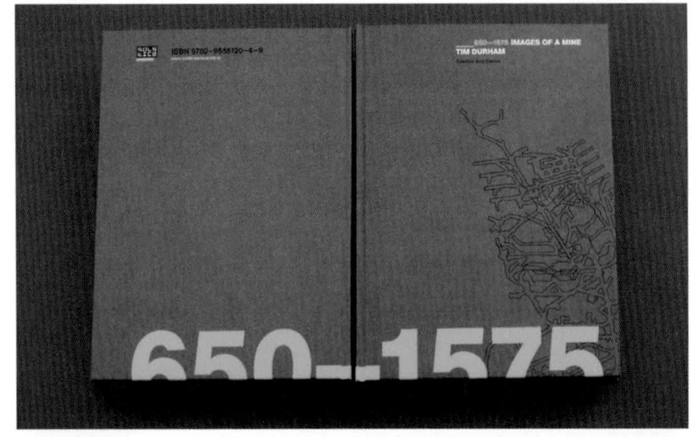

02

▶ 01

设计者：Atelier David Smith
作品名：*10th Anniversary Book*
“消防站艺术家工作室”

▶ 02

设计者：Atelier David Smith
作品名：*650-1575: Images of a Mine*
摄影：Tim Durham
“至日艺术中心”

▶ 03

设计者：Atelier David Smith
作品名：*Raw Alternatives*
全国大学生设计大赛海报

Designers Choice

RAW
ALTERNATIV
ES NATIONAL
STUDENT
DESIGNCOM
PETITION

www.rawalternatives.com

www.terreus.com

The competition is open to all final year and penultimate year students on a registered graphic design or visual communications course in the Republic of Ireland and Northern Ireland.

The brief is available online or from Heads of Departments and Course Tutors. Students can register their interest at www.rawalternatives.com.

The competition is sponsored by Designers Choice, (the specialty division of McNaughton Paper, Ireland) and Terreus® the premium range of paper from Designers Choice.

The competition is supported by the Institute of Designers in Ireland (IDI) and Graphics International Magazine.

The overall winner will receive a cheque for €1000 from the sponsor. The winning student and two runners-up will be awarded certificates of excellence by the judging panel.

Winning entries will be showcased in Graphics International magazine and exhibited online at www.rawalternatives.com

The judges for this years competition are Conor Clarke (Design Factory), Lisa Godson (The Sunday Times) and Tara Clarke (ZINC). Further details available online.

The deadline for entries is Friday 28th March 2003.

[designerschoice]  Terreus® Graphics International

03

06.2.4
数字主角

基于对数字和文字游戏的双关语，“Hörður Lárusson”以年龄和绰号“Hö”为他30岁生日打下烙印。因为派对在冰岛的农村举行，受邀来其生日派对的客人都会收到一枚徽章和一本包含正式邀请函及含有派对相关信息的说明小册子。为此还设计了一面旗帜对派对的地点进行标志。

“Remake”设计工作室为“为了艺术的志愿律师”设计了的另一份年度报告，用前一年的数字逐渐淡出并半透明，而当年的数字则以加深并增亮的方式来呈现年度报告所涉及的两个年度。

“Remake”为“蒂格”工业设计公司设计的宣传手册使用大号的定制数字来突出部门的划分。

04

▶ 04

设计者：Hörður Lárusson
作品名：*3ö*
铭刻设计者的30岁生日派对

▶ 05

设计者：Remake Design
作品名：*Volunteer Lawyers for the Arts*
年度报告封面

▶ 06

设计者：Remake Design
作品名：*Teague*
“蒂格”工业设计公司的宣传手册

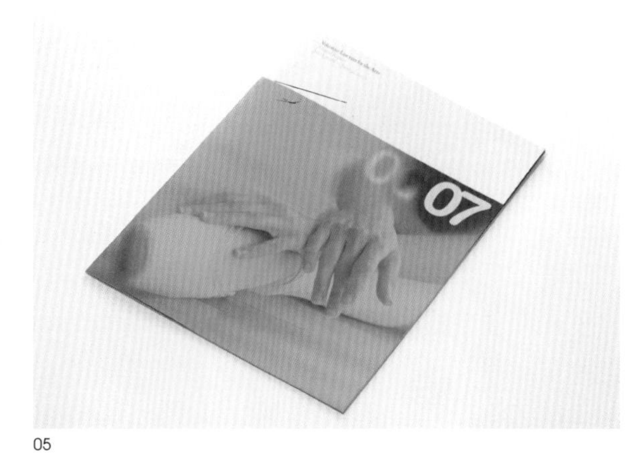
07

05

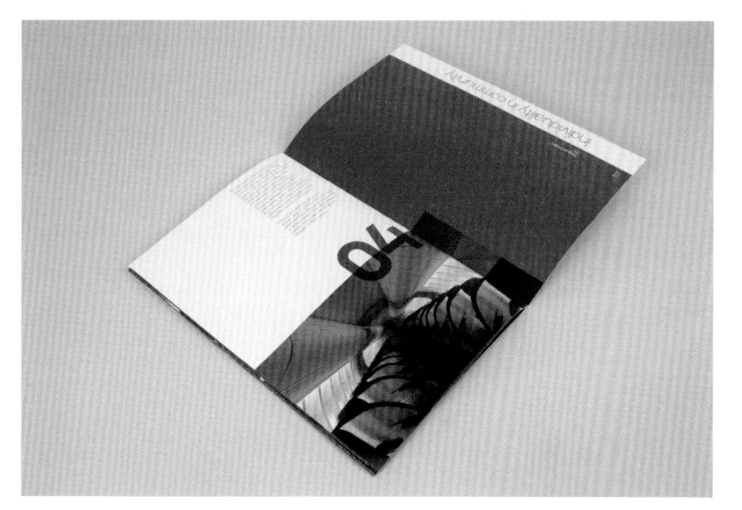
Individuality in community
01

02
Horizon up
Teague insight
02
06
07
In our opinion, the most successful cabin architecture is a harmonious relationship between the airline brand and the aircraft manufacturer's expressions. The latter being subordinate to the former. Like a great gallery or museum space, we consider the cabin envelope as a platform or blank canvas that fully realizes and enables the airline brand's message and expression.

06

06.2.4
数字主角

值传奇设计人物马西莫·维格尼利（Massimo Vignelli）80岁生日之际，“Bibliothèque”在南非设计会议上向他赠送了《马西莫生日快乐》海报。海报再次沿用图标性的365天万年历（31块螺旋装订的面板，每天转动一次来显示日期）应用于数字“80”。目前海报悬挂在纽约“维格尼利协会”的墙上。

房地产投资方“顾问集团”以“邦奇”设计公司（Bunch Design）为其设计的新型年度计划表，作为2010年新年的开端。通过丝网印刷制造出的200份计划表是由年度数字有趣组合而成。

“五角设计”为陶瓷商兼设计师艾米丽·约翰逊设计了标志和包装。艾米丽的祖先“约翰兄弟”1882年创立了这个企业，她是遵循祖先脚步的第五代传人。当艾米丽与她的父亲成立一家新的制陶公司：“1882有限责任公司”时，是他们对祖先留下遗产的致敬。

07

08

▸ 07

设计者：Bibilothèque
作品名：*Happy Birthday Massimo*
为设计界传奇人物设计的海报

▸ 08

设计者：Bunch Design
作品名：*Consult ant Group 2010 Limited edition screen -printed planner*
“顾问集团”2010年限量版丝网印刷年度计划表

▸ 09

设计者：Pentagram
作品名：*1882*
标志和包装

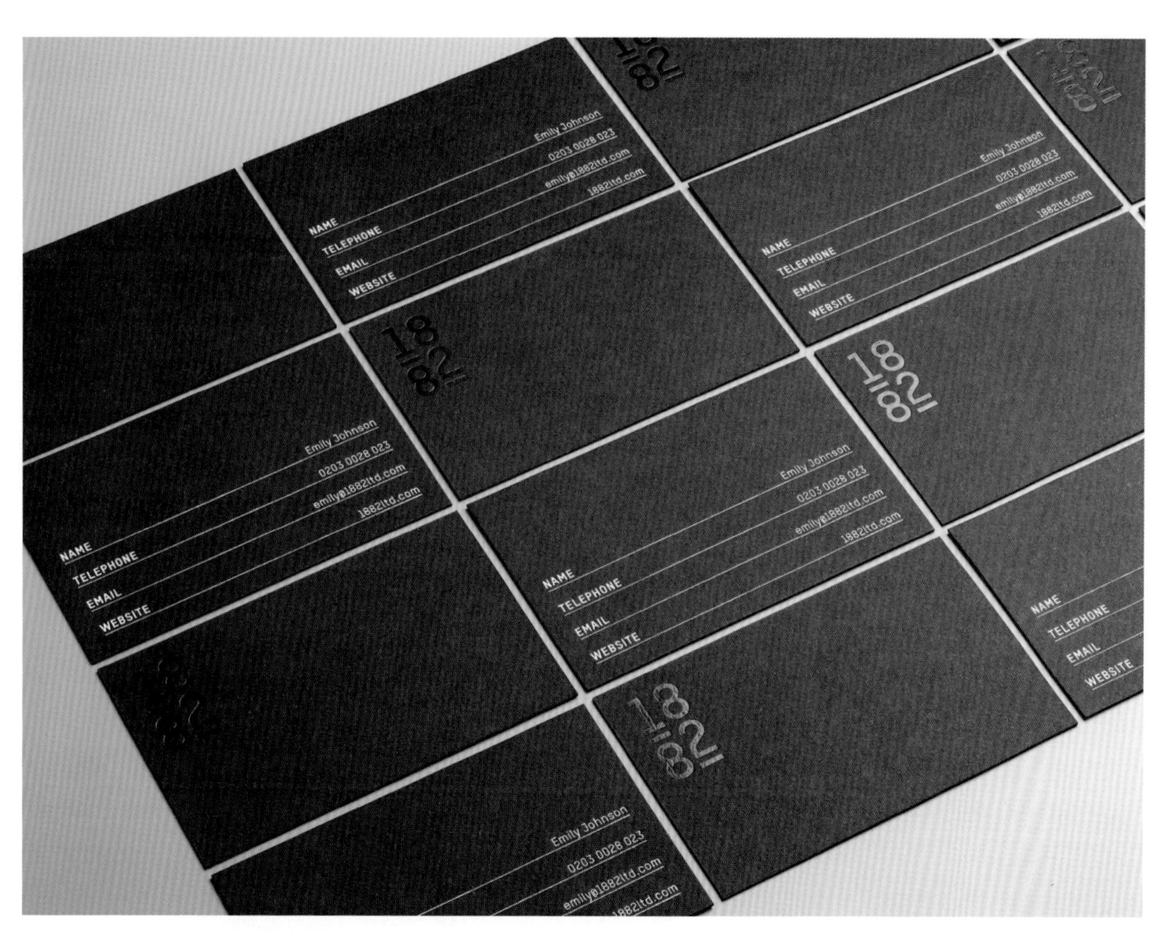
NAME
TELEPHONE
EMAIL
WEBSITE
Emily Johnson
0203 0028 023
emily@1882ltd.com
1882ltd.com
18
82

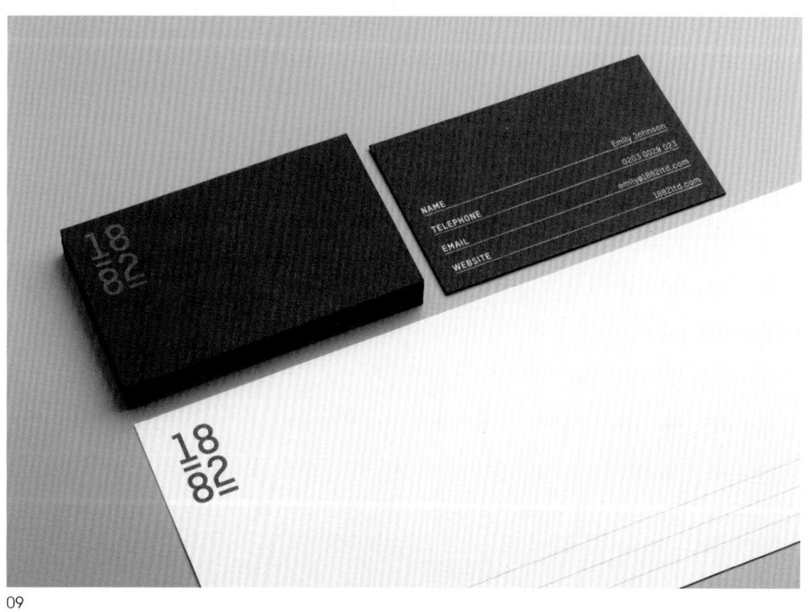
NAME
TELEPHONE
EMAIL
WEBSITE
Emily Johnson
0203 0028 023
emily@1882ltd.com
1882ltd.com
18
82

18
82

09

06.2.5
华丽的排版

《奇克》杂志的展开页面由“曼尼奇联合公司”设计，版式设计由摆放数字“3 000cc”构成。数字“0”中间强大的椭圆空间形成垂直面与水平面间的强烈对照。

法国书籍设计师和印刷商罗伯特·马桑的专著其特色在于用数字组成来表示书中的6个章首页。每章的开场重复同样的构成，章节数字用白色突出。

01

▸ 01

设计者：Marnich Associates
作品名：*Chic magazine*
时尚杂志的展开页面

▸ 02

设计者：Struktur Design
作品名：*Massin Chapter Openers*
著名法国平面设计师的专著

folio

Le théâtre parlé
1 2 3 5 4 6
« [...] une pièce de théâtre risque de n'être qu'un corps sans vie si le typographe, entendant bien son rôle, ne s'emploie à donner à ce lecteur le sentiment que son fauteuil est celui d'un théâtre¹. »

La direction artistique
1 2 3 5 4 6
« Si vous me donnez carte blanche, je me chargerai de la typographie de tous les livres qui paraissent sous votre marque¹. »

02

fantatique!

06.2.5
华丽的排版

莱昂纳多·索诺里（Leonardo Sonnoli's）（意大利）为“Petroltecnica”的广告宣传设计的手绘数字，使用看似抽象的曲线元素来组成数字25。

数字“150”在粗细程度上的极大反差形成漂亮的排版。

这里显示的是《Elephant》杂志第五期包含的四种定制标题字体及一种定制数字设计精选。

03

04

▸ 03

设计者：Leonardo Sonnoli
作品名：*25*
Petroltecnica Terra Therapy

▸ 04

设计者：Leonardo Sonnoli
作品名：*150*
手绘字体

▸ 05

设计者：Studio8 design
作品名：*Elephant magazine*
标题页及内容

COLOPHON
DISTRIBUTION

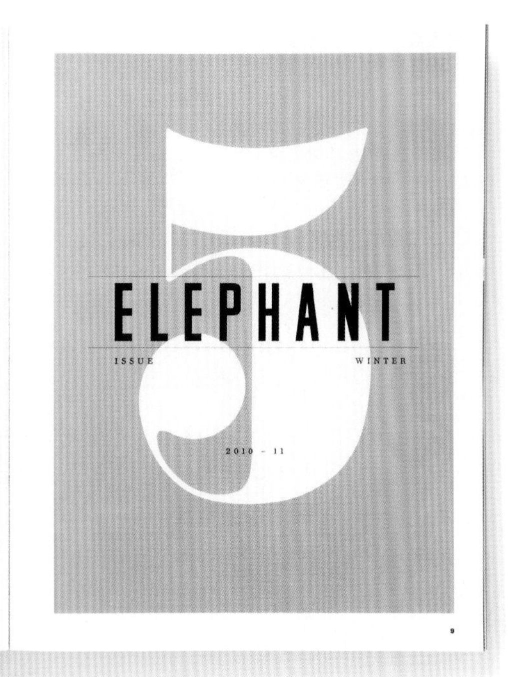
ELEPHANT
ISSUE
WINTER
2010 - 11

MONICA NARANJO URIBE
WHERE HERE MEETS THERE
JOSEPH HART
IN THE BROADEST SENSE
STUDIO & OTHER VISITS
LUDOVICA GIOSCIA
COLOURFUL HEDONISM
GUTO LACAZ
THE PHYSICS OF JUGGLING
MAYA GOLD
THE VIEW FROM ABOVE
JEFF NISHINAKA
PAPER DRAGONS
THE ART OF COLLAGE
RESEARCH
FEATURING WORK BY & INTERVIEWS WITH:
LOU BEACH
CLESS
SIMON COOK
LIAM CROCKARD
VALERO DOVAL
JOSEBA ELORZA
JAMES GALLAGHER
SILJA GOETZ
EVA LAKE
SAM LUBICZ
CLARA MATA
BRION NUNDA ROSCH
JULIEN PACAUD
MASHA RUMYANTSEVA
BRETT RYDER
MARIO WAGNER
JACOB WHIBLEY

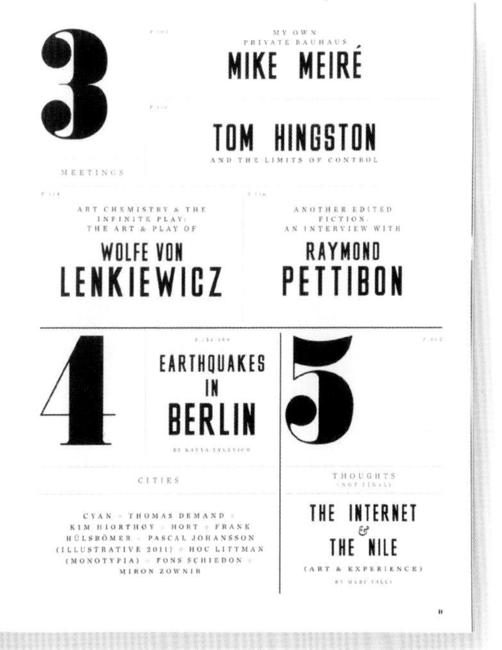
MY OWN PRIVATE BAUHAUS
MIKE MEIRÉ
TOM HINGSTON
AND THE LIMITS OF CONTROL
MEETINGS
ART CHEMISTRY & THE INFINITE PLAY: THE ART & PLAY OF
WOLFE VON LENKIEWICZ
ANOTHER EDITED FICTION: AN INTERVIEW WITH
RAYMOND PETTIBON
EARTHQUAKES IN BERLIN
CITIES
CYAN · THOMAS DEMAND · KIM HIORTHØY · HORT · FRANK HÜLSBÖMER · PASCAL JOHANSSON (ILLUSTRATIVE 2011) · HOC LITTMAN (MONOTYPIA) · FONS SCHIEDON · MIRON ZOWNIR
THOUGHTS
THE INTERNET & THE NILE
(ART & EXPERIENCE)

05

06.2.6
不成比例

普利尔·维切利·克雷默斯（Prill Vieceli Cremers）的书籍《Stadion Letzigrund Zürich》是为庆祝新体育场的启用而创作的，使用的页码数大小各不相同，从占一半页面到更适中的比例，成为页面设计中的关键部分。大号数字的使用与所有运动项目的种类密切相关，看起来就像固定在运动员背后或大型记分板上的数字一样。

01
设计者：Prill Vieceli Cremers
作品名：*Stadion Letzigrund Zürich*
书籍设计

044

Der Ring schliesst sich. Arbeiten am Stadiondach und Betonieren der Sitzstufen, Dezember 2006 (Fotos Helmut Wachter)

The ring closes. Work on the roof of the stadium and pouring the concrete for the seating areas, December 2006 (Photos Helmut Wachter)

056

Rudolf Hirt

Implenia Generalunternehmung AG, Dietlikon; Gesamtprojektleiter Stadion Letzigrund

Implenia General Contractors Ltd., Dietlikon, General project manager for the Letzigrund Stadium

Foto/Photo: Helmut Wachter

Der Macher

Der Letzigrund ist wahrscheinlich das anspruchsvollste Projekt, das ich bis jetzt geleitet habe, obwohl ich schon dreissig Jahre Erfahrung im In- und Ausland mitbringe. Beim Bau dieses Stadions gab es sehr wenig Routine; man musste jedes Detail neu durchdenken und möglichst früh versuchen, alle Risiken vorauszusehen und Sicherungsnetze einzubauen, oft sogar mehrfache. Im Letzigrund durfte nichts schiefgehen, denn es gab keine Hauptprobe: Am 7. September 2007 musste alles auf Anhieb funktionieren, dafür garantierte die Implenia Generalunternehmung AG als Totalunternehmerin.

Wir teilten das gesamte Baufeld in mehrere Einzelbaustellen mit je eigener Leitung auf, die parallel nebeneinander arbeiten und koordiniert werden mussten. Natürlich war auch der Bauablauf in Etappen und Phasenpläne gegliedert, in Abhängigkeit von der Logistik: Hier mitten in der Stadt war die Zu- und Wegfahrt der Lastwagen ein Engpass, den man sehr genau planen musste.

Früh setzten wir klare Prioritäten. Wir suchten zuerst nach den kritischen Punkten im Bauprozess, das war hier der Stahlbau. Nicht nur wegen der hohen Anforderungen an die Präzision, sondern auch wegen des Zeitaufwandes: Nach dem Einbetonieren der Zugstützen muss man 28 Tage warten, bis es weitergehen kann. Daher hatte der Stahlbau jederzeit den Vortritt, die anderen Baustellen durften die Stahlbaumontage nicht behindern. Dann, gegen Schluss der Bauzeit, hatte der Elektrounternehmer Priorität. Von ihm hing es ab, ob an der Eröffnung alles funktionieren würde: Drehkreuze, Beleuchtung, Audioanlage – also das, was der Besucher wahrnimmt. Allein im Dach verlaufen 65 Kilometer Kabel!

Im Letzigrund arbeiteten wir hauptsächlich mit Firmen zusammen, die wir aus bewährter Zusammenarbeit kennen. Rund 36 Prozent der Aufträge gingen an unsere eigene Bauunternehmung, die Implenia Bau AG. Bauen ist Teamarbeit, und der Teamgedanke stand auf dieser Baustelle im Zentrum. Im Letzigrund haben wir durch gute Kommunikation eine positive Spirale auslösen können. Alle sind stolz, an diesem Werk mitgewirkt und die Verantwortung mitgetragen zu haben.

The Doer

The Letzigrund Stadium is probably the most ambitious project I have led thus far, even though I bring with me thirty years of experience here and abroad. Little about building this stadium was routine; every detail had to be thought through, and all of the risks had to be foreseen as far in advance as possible and multiple safety nets set up. Nothing could go wrong with the Letzigrund, because there was no dress rehearsal: it all had to function from the start on 7 September 2007; as the company responsible for planning and contracting, Implenia General Contractors Ltd., guaranteed that.

We divided the construction area into several building sites, each with its own management team, but they had to work in parallel in a coordinated way. The construction process was of course also divided into stages and which depend on logistics: the arrival and departure of lorries in the middle of the city was a potential bottleneck that had to be precisely planned.

We set priorities early on. First, we determined the critical points in the building process, such as the steel construction. It is not just the precision necessary but also the time required: a twenty-eight-day wait is required after tensile supports have been set in concrete. For that reason, steel construction always had priority; the other construction sites could not slow the assembly of the steel construction. Then, as the construction phase was coming to a close, the electricians had priority. It was up to them whether everything would work for the opening: turnstiles, lighting, audio system – everything the visitors see. There are sixty-five kilometres of cable in the roof alone!

At the Letzigrund we've been working primarily with companies we know from previous collaborations. Approximately thirty-six percent of the contracts went to our own construction company, Implenia Bau AG. Construction is teamwork, and the team was the focus at this site. At the Letzigrund, we were able to trigger a positive chain reaction through good communication. Everyone was proud to be part of this project, and they shared the responsibility.

Judit Solt

Kraftvolle Dynamik

Das Stadion Letzigrund aus architektonischer Sicht

Judit Solt

A Powerful Dynamic

The Letzigrund Stadium from an Architectural Perspective

057

Der Letzigrund hat etwas Surreales. Trotz seiner Grösse wirkt das Gebäude keineswegs überdimensioniert, scheinbar mühelos bringt es die gegensätzlichen Anforderungen von Sport und Showbusiness in Einklang, verblüffend leicht wirkt das elegant geschwungene Tribünendach. Hinter der einfachen, prägnanten Form verbirgt sich indes ein hoch komplexer Entwurf. Formale, funktionale und konstruktive Aspekte verbinden sich zu einer faszinierenden Einheit.

«Als ich hinein trat, mehr noch aber, als ich oben auf dem Rande umher ging, schien es mir seltsam, etwas Grosses und doch eigentlich nichts zu sehen. Auch will es nicht leer gesehen sein, sondern ganz voll von Menschen. [...] Denn eigentlich ist so ein Amphitheater recht gemacht, dem Volk mit sich selbst zu imponieren, das Volk mit sich selbst zum Besten zu haben. Dieses allgemeine Bedürfnis zu befriedigen, ist hier die Aufgabe des Architekten. Er bereitet einen solchen Krater durch Kunst, so einfach als nur möglich, damit dessen Zierrat das Volk selbst werde. [...] Die Simplizität des Ovals ist jedem Auge auf die angenehmste Weise fühlbar, und jeder Kopf dient zum Masse, wie ungeheuer das Ganze sei.»

Johann Wolfgang von Goethe[1]

The Letzigrund has something surreal about it. Despite its scale, the building seems by no means outsized; seemingly without effort, it brings together the contradictory demands of sport and show business; the elegantly curved roof over the grandstands looks astonishingly light. Behind the simple, concise form, however, a highly complex design is concealed. Formal, functional and constructional aspects combine into a fascinating unity.

'As I entered it, and even more when I wandered about on its highest rim, I had the peculiar feeling that, grand as it was, I was looking at nothing. It ought not to be seen empty but packed with human beings ... Such an amphitheatre, in fact, is properly designed to impress the people with itself, to make them feel at their best ... To satisfy this universal need is the architect's task. By his art he creates as plain a crater as possible and the public itself supplies its decoration ... The simplicity of the oval is felt by everyone to be the most pleasing shape to the eye, and each head serves as a measure for the tremendous scale of the whole.'

Johann Wolfgang von Goethe[1]

06.2.7
数字图像

日本邮购公司“爱速客乐”24小时内在全国范围的宅配货物超过40 000件。其服务项目覆盖从家具、文具、计算机到食品及日用品的各类产品。“斯德哥尔摩设计实验室”（Stockholm Design Lab）重新设计了该范围内的50到60种重要产品。他们的目的在于以更明确和更生动的方式来区分产品：明晰代表更多的销售。关键在于寻找和运用客户第一眼就能理解的标志。例如电池的型号（1~4）及图形数字“6”和“7”用来表明线纹信纸的有效行距为6~7 mm。

赫尔穆特・舒密特（Hlelmut Schmid）为“Hinex”设计的一系列小册子封面让大号数字序列发挥主导作用：数字及所有其他字体均设置为25度角，以在封面上创造出动态拉伸的效果。

01

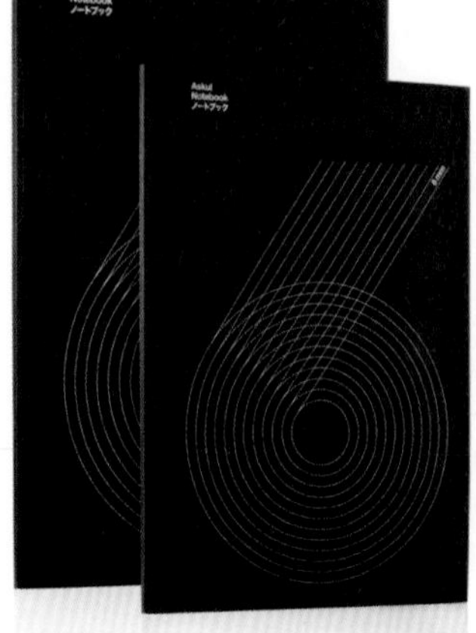

02

▸ 01

设计者：Stockholm Design Lab
作品名：*Askul*
线纹笔记本

▸ 02

设计者：Stockholm Design Lab
作品名：*Askul*
电池设计

▸ 03

设计者：Helmut Schmid
作品名：*Hinex*
小册子封面

大塚製薬株式会社
1
生理性
○自然食主体でニュータイプの
Enteral Hyperalimentationです
○栄養素のバランスがとれた組成です
○消化吸収され易く、消化管粘膜への刺激が少ない
Enteral Hyperalimentation
ハイネックス
HINEX

大塚製薬株式会社
2
有効性
○高カロリー補給が可能です
○タンパクの利用がスムースです
○脂質代謝に好影響を与えます
Enteral Hyperalimentation
ハイネックス
HINEX

大塚製薬株式会社
3
安性性
○長期投与が可能です
○長期保存が可能です
○副作用の頻度が従来製品と比較して非常に少ない
Enteral Hyperalimentation
ハイネックス
HINEX

大塚製薬株式会社
4
便利性
○優れた流動性を有しています
○調製が容易です
○保存が容易です
Enteral Hyperalimentation
ハイネックス
HINEX

03

06.2.7
数字图像

"斯德哥尔摩设计实验室"（Stockholm Design Lab）为日本邮购公司"Askul"设计的更多包装设计实例表明，设计者的首要目的在于创造良好的功能性设计，同时他们也希望去掉所有不必要的细节，并考虑增加趣味元素。各类办公文具的包装让人回想起20世纪六七十年代的药品包装，蕴含清晰、内敛及国际化设计的风格。

USB记忆棒的设计再次简化为一个色块板和一个大号清晰的数字，数字用来说明每个记忆棒所包含的字节空间。

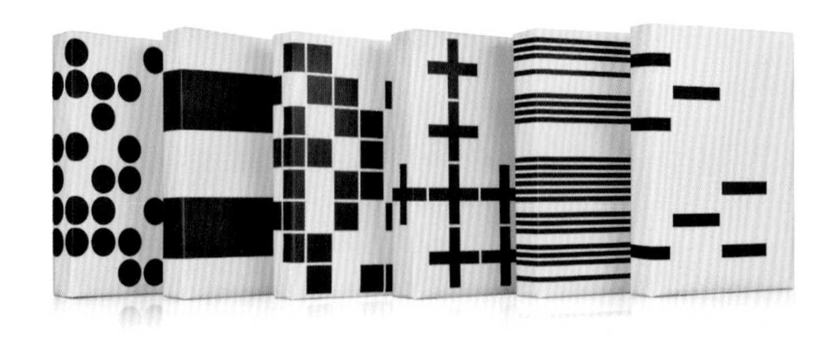

05

06

07

▸ 04
▸ 05

设计者：Stockholm Design Lab
作品名：*Askul*
不同的文具用品

▸ 06
▸ 07

设计者：Stockholm Design Lab
作品名：*Askul*
1, 2, 4 及8GB USB 记忆棒

修正テープ
6 mm × 15 m
10個入
6mm
修正テープ
5 mm × 15 m
10個入
5mm
Askul
Index tabs
220シート
22シート×10袋・3520片
インデックスラベル青 (小)
Askul
Index tabs
220シート
22シート×10袋・3520片
インデックスラベル赤 (小)
修正テープ
4.2mm × 15 m
10個入
4.2mm
Askul
オリジナルふせん
のりワイド
75×75mm

04

06.2.8
排序

《平面设计要素》是一本简单的分步指南手册，它用来帮助设计者熟悉设计软件程序中的基础知识。使用大号的数字来描述书本的8个部分。内容列扩展为4页，每个章节采用超大型的数字，之后在封面设计中对章节数字进行了重叠。

01

设计者：Struktur Design
作品名：*Graphic Design Essentials*
书籍封面及目录

1

Introduction

2

The Elements of Design

3

Typography

4

Images

5

Layout

6

Logo Design

7

Visual Themes

8

Information

06.3 表达

06.3.1 线性

“Commercial Type”创建的特殊标题字体被用于由“Working Format”公司设计的《彭博商业周刊》上。字体由同心的平行线构成，使用大号字体且超出页面。即使在这种尺寸上，字体也仍保持浅灰色调，因而不会对页面上的编辑内容造成压制或与之相冲突。

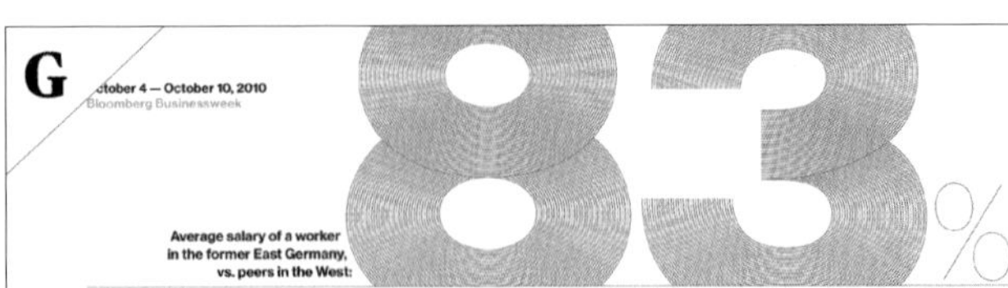

G ctober 4 — October 10, 2010
Bloomberg Businessweek

Average salary of a worker in the former East Germany, vs. peers in the West: 83%

cating Merkel to rescue the system not because she loves Europe, but because Germany is its leading beneficiary.

While Kohl battled to keep German and European interests in perfect alignment, he was also the consummate domestic political opportunist. He dodged and weaved for months in 1990 before recognizing Poland's border with Germany as inviolable–the same kind of fence-straddling Merkel was accused of when Greece pleaded for help. And by the end of his tenure in 1998, Kohl, with his treasury buckling under the cost of subsidizing the East, was making idle threats to stop paying into the EU budget.

So when critics–including many from her Christian Democratic base–expressed outrage over Merkel's handling

as the EU grew from 6 countries in 1957 to 15 in 1995, and it worked for the creation of the euro in 1999. But things are different in today's larger bloc. Defeats for German-French projects used to be rare; now they are increasingly common, as when other governments rejected France and Germany's argument for a European financial transaction tax on Sept. 7. "For a long time the German-French relationship has been much more appearance than reality," says Günter Verheugen, 66, Germany's EU commissioner from 1999 to 2010. "The days are gone when a German-French directorate could set the course of European integration."

It's not only a numbers game. German and French interests have been drifting apart since the end of the Cold War. As

abroad. Some 64 percent of that investment is in the EU, sustaining 2.9 million jobs, according to Deutsche Bundesbank data. "The export model directly benefits other countries," says Heleen Mees, a researcher at the Erasmus School of Economics in Rotterdam. "It's very worthwhile for those countries to have German factories and technologies."

Merkel has won on one point. As the EU tries to strengthen the euro region's financial management, the onus is on weaker countries to improve how they do things, not on Germany to abandon fiscal rigor and loosen the lid on wages. Yet that's about the only point she has taken so far. German calls for harsher sanctions on deficit offenders face resistance, and German pleas for an "order-

G ctober 4 — October 10, 2010
Bloomberg Businessweek

Known number of people between 1961 and 1989 who died attempting to cross the Berlin Wall: 136

omy Minister Sven Morlok. Two key figures in the transformation were Kurt Biedenkopf, now 80, the Christian Democrat who served as premier from 1990 to 2002, and the late Kajo Schommer, who served as economy minister during the same formative period. Together they developed a "lighthouse principle" in which industries with high potential–especially autos and microelectronics–got most of the state aid.

"The state's political and business leaders boosted the self-confidence of the Saxon people and convinced them their fate was in their own hands," says Joachim Ragnitz, managing director of the Ifo Institute for Economic Research in Dresden. "It was like nation-building, and it worked. They got people to believe that they really had to be best at whatever

and his team," says Jens Drews, Globalfoundries' director of government relations. "They were visionaries."

With 4.2 million people, Saxony sits in an eastern nook of Germany, with the Czech Republic to the south and Poland to the east. Twenty years after reunification, about 6,000 companies have invested €27 billion ($35 billion) in the state, which is smaller than New Jersey. East Germany's anticommunist mass protests had their genesis in the Saxon city of Leipzig in 1989, paving the way for demonstrations that swept the hard-line state before it imploded that year. As a result, Saxon voters have not elected any former East German communist to power as part of any coalition government.

The local government has taken market-friendly steps that would be unthink-

cleaner and more prosperous, the church towers bigger and higher."

Porsche's plant is a symbol of the new Saxony, employing 640 people on the site of a former Soviet military training base. It churns out 250 Cayennes and 100 Panameras a day. The meticulous attention to detail that makes Saxons good at building Porsches is also an asset for producing luxury products such as watches and fine wine. A. Lange & Söhne, the watchmaker established in 1845 in the village of Glashütte near the Czech border, was expropriated by the communists in 1945 and reestablished in December 1990 by Walter Lange, a direct descendant of the founder. Jerzy Schaper, Lange's CEO, says watches made from gold and platinum–which range in price from €13,000 to €390,000–are selling especially well in Asia. Georg

60

the challenges of a globalized world. Next, Germany and its neighbors imagined a unified Europe, and the country paid for it with spending power when the strong deutsche mark was retired for a then-weaker euro. Now Germans are paying again, to prevent Greece from pulling apart the European Union. It's a fresh reminder that togetherness has its costs–though 20 years later, the benefits of reunification speak for themselves.

6%

Percent of Germans who own stocks

Making What?
The *Heiss* List, page 70

What Drives the German Economy?
New Rules and Old Companies, page 72

Who's Earning, and Where?
The Workscape, page 76

Why Don't Germans Like Stocks?
Chronic Bubble Phobia, page 77

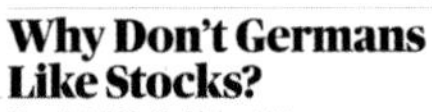

What Makes Saxony Run?
The Rebirth of Dresden, page 77

How Did Siemens Go From Scandal to Success?
"Never Miss a Good Crisis," page 79

How Does It Feel to Be a Turk in Germany?
A Merger in the Shadows, page 80

Plus: Essential info on doing business in Germany —page 82

▸ 01

设计者：Working Format / Commercial Type
作品名：*Bloomberg Businessweek*
定制的字体设计

October 4 — October 10, 2010
Bloomberg Businessweek

Number of Olympic medals East Germany won, from 1968 to 1988. West Germany won 253.

We wanted
porate cul-
s Yerli, 40,
nployees at
s including
company's
German el-
d planning,
quality."
n residents
shrinking,
has reason
nic minor-
ummer in-
out wheth-
mmigrants
r" and sin-
ty as most
se incendi-

When soccer player Mesut Ozil scored for Germany at the 2010 World Cup in South Africa, fans back home didn't care where his parents were from. Cem Ozdemir, whose Turkish-born mother once worked as a seamstress in southern Germany, became the nation's first federal lawmaker from an immigrant family in 1994 and now heads the opposition Green Party, which is polling at a record high.

Turkish-Germans run some 80,000 businesses that employ, on average, five people. Still, they're only about half as likely as Germans to start companies, according to the Turkish Community in Germany, an advocacy group.

Driving anti-immigrant sentiment is a large, unassimilated Turkish underclass community has the lowest share of high school graduates (14 percent) and an unemployment rate of 23 percent, according to a 2009 study by the Berlin Institute for Population and Development. That compares with a 7.6 percent jobless rate for Germany in August. Turks and their children are twice as likely as Germans to wind up on welfare, and in Berlin they top the police tally of gang criminals.

"Yes, there is discrimination" against non-German job seekers, says Nihat Sorgec, head of a vocational school in Berlin's heavily immigrant Kreuzberg neighborhood. "But whining and feeling victimized is not the best way to fight this. You need to move on and say, 'These are my qualifications, I can do this job better than others.'" His wish is "We can't waste any talent." Asked what non-Germans need to do to get ahead, politicians from Merkel on down reply: Learn German. "Whoever lives among us also has to be ready to integrate into society, learn the language, participate in school," Merkel said in August. "There's a lot to do in that respect."

Turkish community leaders say Germans need to be more welcoming. Only a decade ago politicians were still debating whether Germany was an "immigration country" at all. In 2000, after the Social Democratic government of Gerhard Schröder changed the country's citizenship law to reflect growing diversity, a leader of Merkel's Christian Democrats called for a "guiding culture" of Germanness. Another complicating factor is Eu-

1989

tober 10, 2010
nessweek

20

ker, but it was also the start
uring which Germans strug-
-not just to each other but
ple in Europe and their place
order. "It's like a marriage,"
Hohenthal, a former political
nt for the German daily *Die*
you start out, you are very
, and even though you might
d you aren't sure about ev-
decide to do it. But eventu-
ts in."

for Unification"

e Berlin Wall did not make
evitable. Although nominally
, the two German states had
xisted under the controlling
he four victorious World War
U.S., Great Britain, France,
Union. As of 1989, hundreds
of foreign troops, including
million members of the Red
ill stationed on German soil.
leading up to Nov. 9, discon-
Communist regime in East

The Reichstag with Berlin Wall

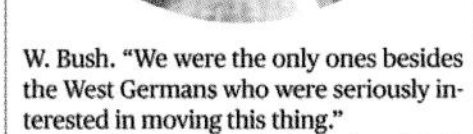

W. Bush. "We were the only ones besides the West Germans who were seriously interested in moving this thing."

The West Germans had doubts of their own. Helmut Kohl, the conservative Chancellor of the Federal Republic, told aides unification would take at least five years, if it happened at all. Other West German politicians were even more skeptical. "A big part of the political class here did not be-

opening, Kohl made a trip to Dresden, the bombed-out city that was still a part of the East. Throngs came out to cheer him and plead for German unity. "Kohl was not a great speaker, but he had a tremendous feeling for the people," says von Hohenthal, who covered Kohl's trip for *Frankfurter Allgemeine Zeitung*, Germany's largest newspaper. "I remember looking at the faces of the people—they all were wearing black, red, and gold and chanting, 'Helmut, Helmut!' You could just sense that the East Germans were absolutely longing—screaming—for unification."

By the start of 1990 the Kohl government seized the opportunity and pushed for a rapid, one-state solution. The Soviets presented the biggest dilemma: Letting go of the GDR, Moscow's most important satellite in Central Europe, would be tantamount to conceding defeat in the Cold War. To exact agreement from the Soviets, Kohl turned to Bush, who persuaded Gorbachev to accede to unification and to accept the incorporation of a unified Germany into NATO rather

GERARD MALIE/AFP/GETTY IMAGES (2)

ANIZATION OF MOTOR VEHICLE MANUFACTURERS; FEDERAL STATISTICS BUREAU

tails of how to unite the two states formally and remove foreign troops were worked out through the "2+4" negotiations—involving representatives of the two Germanys, plus those from the four occupying powers—a process devised by Washington to limit the circle of actors with a say over Germany's future. The final treaty reestablishing German sovereignty was drafted by the end of the summer and signed in Moscow on Sept. 12, 1990, astonishingly quick by the glacial standards of international diplomacy.

The negotiators had glossed over the consequences for the citizens of the GDR, many of whom were unprepared for the coming transformation of their way of life. Yet it's difficult to see how it could have happened any other way. Within a year of the Moscow treaty, Gorbachev was pushed out of power by hardliners in the Kremlin. The Soviet empire collapsed. Negotiating with more mercurial figures in Moscow, let alone with the leaders of 15 newly independent former Soviet republics, might well have proved impossible for

The Reichstag, post-Wall

living conditions in much
East Germany show few sig
ment. With the disappear
nal borders, some 2 milli
mans moved west, sapping
what little intellectual capit
cal knowhow it had left. In
once heavily subsidized by
nists collapsed. At least
early moves, to introduce

06.3.2 焦点

“艾格斯+迪亚柏”（Eggers + Diaper）在过去10年里为柏林犹太博物馆设计周年纪念晚宴邀请函。其每年遇到的挑战是怎样将纪念日数字与标题相结合。大多数年度里，数字代替标题中的一个字母或成为其一部分。灰色与红色的色彩模式保持不变，而排版上的处理则每年都不同。

《第二届芝加哥国际海报双年展》的海报，“五角设计”使用数字2作为设计的主焦点。从一端看去，呈螺旋形的白色线条建立在卷起的海报之上。

01

▸ 01

设计者：Eggers + Diaper
作品名：*Anniversary Dinner Invitations*
柏林犹太博物馆

▸ 02

设计者：Pentagram
作品名：*Second Chicago International Poster Biennial 2010*
官方海报

Second Chicago International Poster Biennial 2010
CHICAGO
POSTER

02

06.3.3
书法

标识“3”是由“五角设计”研发，是为位于萨克斯第五大道三楼知名的百货商店所设计。邀请卡与标志和所有主要设计师品牌名称漂亮地交织在一起，形成优雅的数字“3”。

▸ 01

设计者：Pentagram
作品名：*3 at Saks Fifth Avenue*
《最新设计师系列邀请卡》

Marc Jacobs
Chanel

06.3.4
数字形成的形象

由“gggrafik”工作室设计的24岁生日卡调皮地将数字变成在水里滑行的天鹅。

为庆祝意大利男性杂志《GQ》出版十周年，编者邀请十位国际设计师对封面进行设计。这是莱昂纳多·索诺里（Leoardo Sonnoli）的顽皮设计。

为将在海德尔堡的“俄罗斯迪斯科”举行的五周年纪念派对作宣传，“gggrafik”用一个改造后的锤子和镰刀创作出一份海报，形成数字“5”。

01

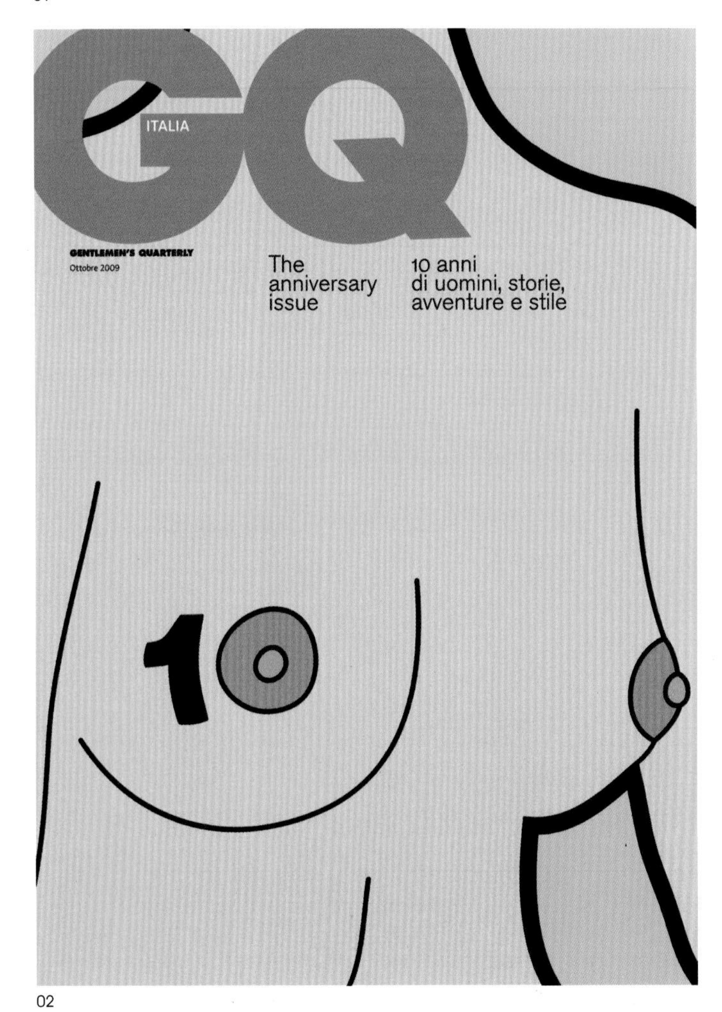

02

▸ 01
设计者：gggrafik
作品名：*24*
生日卡

▸ 02
设计者：Leonardo Sonnoli
作品名：*GQ Italia*
封面设计

▸ 03
设计者：gggrafik
作品名：*Fifth anniversary Russian Disco*
海报

WLADIMIR KAMINER
FUENF JAHRE RUSSENDISKO
ZEHN JAHRE KARLSTORBAHNHOF
MI.7.12. KARLSTORBAHNHOF
20.00 UHR LESUNG/22.00 UHR RUSSENDISKO
Grafik: Götz Gramlich, Druck: Glogner-Druck Heidelberg

03

06.3.5
色彩游戏

多年来，“Ich & Kar”一直为法国里尔的视觉传媒学院（ECV）研发识别系统。这里显示的是几年来特色的诠释，以有趣的排版组合来呈现。颜色与字体保持不变，但数字间的互动却年年不同。

01

02

▸ 01

设计者：Ich &Kar
作品名：*ECV*
2008年标识

▸ 02

设计者：Ich &Kar
作品名：*ECV*
2011年标识

▸ 03

设计者：Ich &Kar
作品名：*ECV*
2009年标识

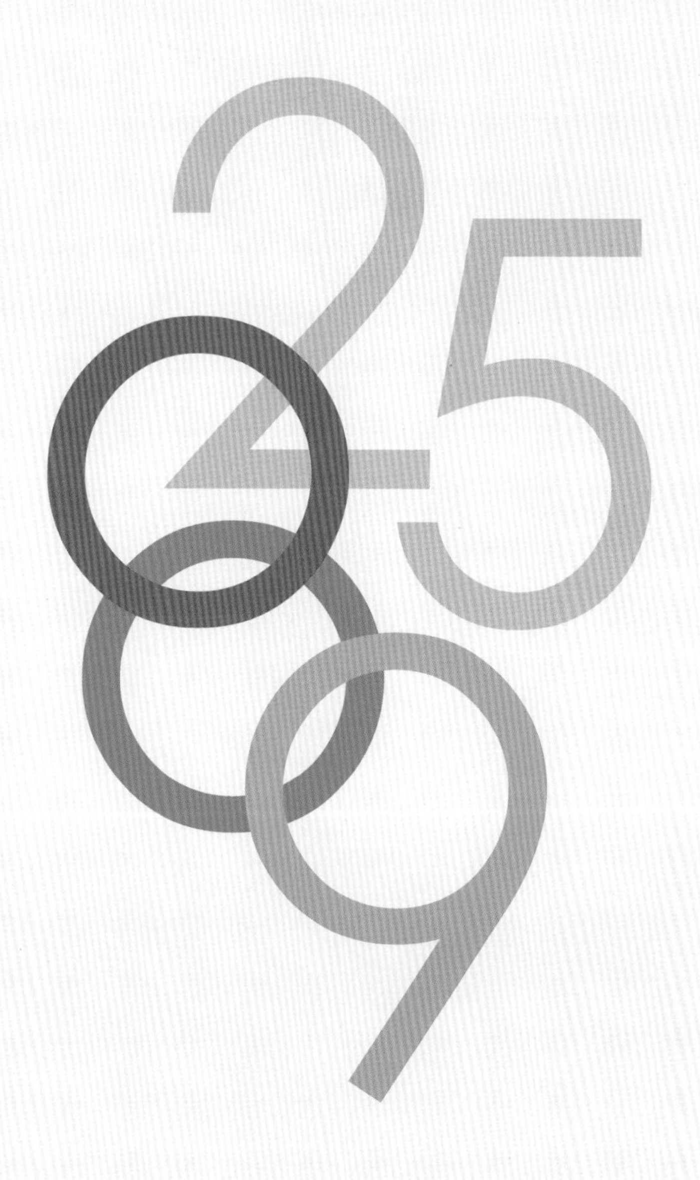

03

07.
增加

07.1 标识系统

07.1.1 数字空间 258
07.1.2 建筑数字 260
07.1.3 数据流 262
07.1.4 单色调 264
07.1.5 与实物一般大小 266
07.1.6 数字导航 268
07.1.7 点阵法 270
07.1.8 漏版阴影 272
07.1.9 透视 274

07.2 超大图形

07.2.1 超大尺寸 276
07.2.2 超越边界 278
07.2.3 全篇幅 284
07.2.4 三维数字 286
07.2.5 数字挡字书板 288

07.1
标识系统

07.1.1
数字空间

爱利街的标志是一套建筑标志系统，这个系统用突出的数字来作为楼层标识。清晰大号的数字在整个建筑物内非常明显，并且与小巧的次要信息相互平衡。

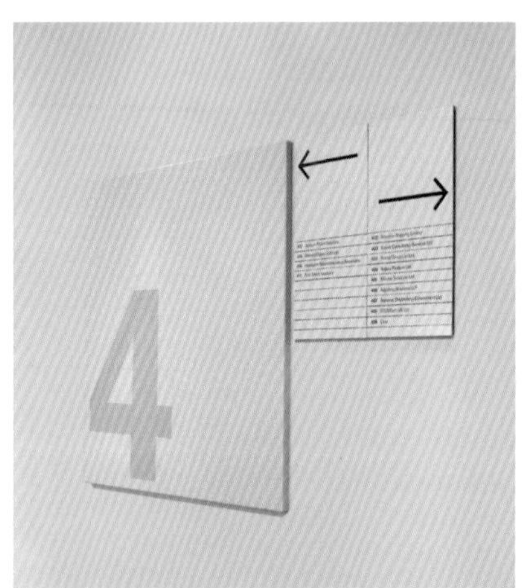

01

设计者：Commercial Art
作品名：*Alie Street Signage*
伦敦金融城公司大楼

1
Alie St.
CITY
LONDON

07.1.2 建筑数字

“五角设计”为“彭博资讯”公司总部设计引导标识、环境标识及媒体设施。该公司提供金融新闻、数据和分析，公司占据曼哈顿东部一座55层塔楼的9层。

建筑物里的寻路标识由数字调整而成。每一层楼由被玻璃包裹得明亮半透明、具有颜色标记的树脂数字来标识。超大型的楼层编号系统延伸到紧急通道，这里极大的数字被漆在墙上和楼梯上。

01

设计者：Pentagram
作品名：*Bloomberg*
导示标识系统

07.1.3
数据流

“彭博资讯”曼哈顿总部第6层有一个区域被称为“连接”，这是一座三层玻璃桥，包含职工和客户的公共通道。在这里，设计师创作出在空间三面滚动的超大新闻链，包括分成四条平行带的媒体演示墙，平衡带从“彭博资讯”现场直播中获取数据。信息流与空间内行人的移动互补，媒体演示墙变幻的色彩使空间整天都处于变化之中。

01
设计者：Pentagram
作品名：*Bloomberg*
实时数据流

07.1.4
单色调

伦敦沃平“大都会码头”一座工业建筑的粗糙白色砖墙上，大型轮廓楼层数字标识在其中并占主要地位。

01
设计者：SEA Design
作品名：*Metropolitan Wharf, Wapping, London*
标识系统

FIRE
EXIT
2

3
3

3

07.1.5
与实物一般大小

“四月”设计工作室为建筑事务所“约翰·麦卡斯拉及其者合伙人”制作的标识设计扩展到公司的新总部。引导标识由紧挨着接待处的一张所有楼层的综合建筑平面图组成，相关数字作为超大图形在每一楼层重复出现。巨大的数字渗入彩色的固体面板，面板上包含会议室及其他设施的位置信息。

01
设计者：April
作品名：*John McAslan + Partners*
大型楼层指示标识

ROOF GARDEN
THIRD FLOOR
STUDIO 3
MEETING ROOM 3
SECOND FLOOR
STUDIO 2

07.1.6
数字导航

比利时“帕斯”科学探险公园的引导标示及导航系统直接由原材料设计而成。大型面板上涂有鲜艳夺目的颜色，颜色与屏幕上的公园虚拟导航地图相关。直接应用于表面的大型版式设计指示着相关区域或楼层。设计成“DIN”字体的数字系统看起来就像来自空间站。

01
设计者：Base Design
作品名：*Pass scientific adventures park*
引导标识与寻路系统

07
06
05

Hangar à expositions

07.1.7
点阵法

斯德哥尔摩“马格努松好酒”设施中所有的酒都储存在现代环境里的最优条件下。除储存服务外，设施为其会员提供专门的酒吧及单独的房间。“斯德哥尔摩设计实验室”与“托马斯·埃里克松设计”一起为位于斯德哥尔摩的储酒设施设计了视觉识别程序及内部设计。简洁的最小化点阵字体遍及储藏空间，以同种方式用模板印刷并应用于各种设施表面。

点阵法也出现在斯堪的纳维亚最大酒店“贝拉天空”的引导标识中。该酒店以“3XN”的构架在哥本哈根的地平线上创造出独特的新形象。两座斜塔高达76.5米。“斯德哥尔摩设计实验室”为其设计了新的标识及引导系统。设计师从建筑物的中获得灵感，设计出了独特的字型。

01

▸ 01

设计者：Stockholm Design Lab
Thomas Eriksson Architects

作品名：*Magnusson Fine Wine*

引导标识与导航系统

▸ 02

设计者：Stockholm Design Lab
Thomas Eriksson Architects

作品名：*Bella Sky*

酒店标识及引导系统

1234567890

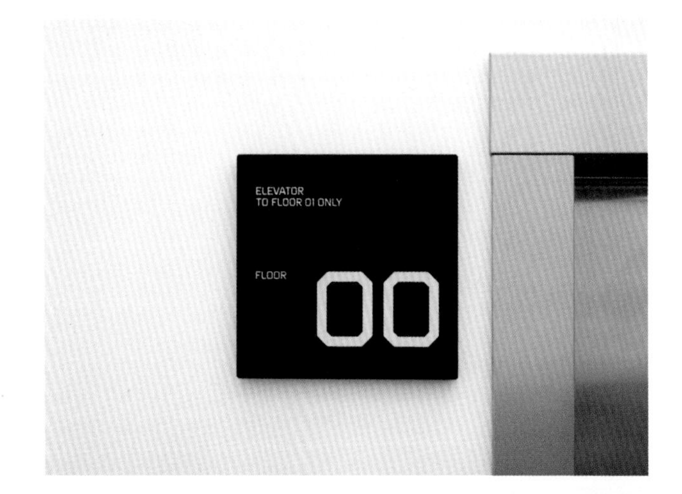

02

07.1.8
漏版阴影

“麦耶斯考夫工作室”使用粗体漏版字体为伦敦传媒学院设计出惊人的原始导示系统。所有数字均为黑色，为繁忙走廊的潜在视觉噪声制造出最强对比。数字的阴影看起来似乎是投射在地板上，这样的超大图形显得更加引人注目，创造出一种三维效果。楼层图案提升了对位置的导航作用，因为下楼的每个人均能看到这些图案。

01
设计者：Studio Myerscough
作品名：*London College of Communications*
导示系统

GALLERY
FROM
TO
BA
ECKERSLEY
0
TUESDAY 09 JUNE
FRIDAY 12 JUNE
BA
LOWER STREET
0
TUESDAY 09 JUNE
WEDNESDAY 10 JUNE
FDA
UPPER STREET
1
THURSDAY 11 JUNE
THURSDAY 18 JUNE
BA
WELL
1
FRIDAY 05 JUNE
FRIDAY 12 JUNE
BA
ATRIUM
1
FRIDAY 05 JUNE
FRIDAY 12 JUNE
BA
MEDIA BLOCK
1
FRIDAY 05 JUNE
FRIDAY 12 JUNE
FDA
0
TUESDAY 09 JUNE
FRIDAY 12 JUNE
BA
OFF SITE
FRIDAY 05 JUNE
SUNDAY 07 JUNE

07.1.9
透视

一条140 m长的走廊连接路德维希堡“Kreissparkasse”银行新大楼各部分。楼层平面及楼梯标识的透视图都非常失真，以至于只能从唯一的点看过去才能认出它们，否则它们将变形成一种自由形式的游戏。

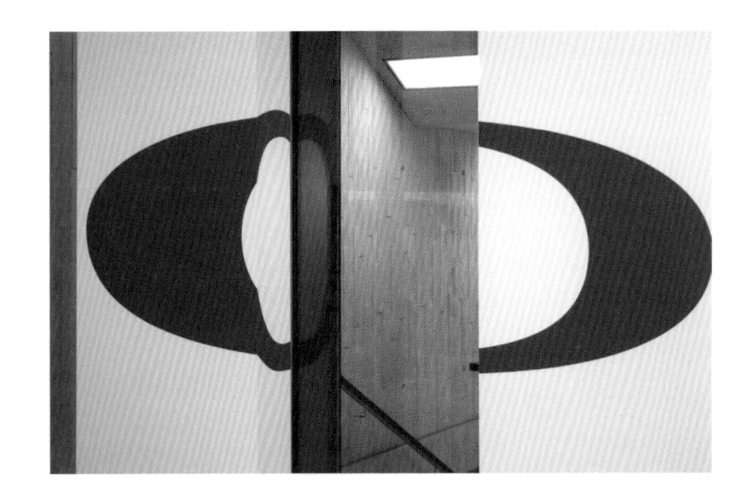

01

设计者：L2M3 Kommunikationsdesign
作品名：*Kreissparkasse, Ludwigsburg*
三维变形标识系统

C2

B2

07.2
超大图形

07.2.1
超大尺寸

位于伦敦中心位置的“巴比肯艺术中心”，其导航系统的设计总是具有挑战性。这一引导标识及导航系统与20世纪60年代原始的混凝土建筑完美和谐地融合在一起。在各处使用强烈的橙黄色，与“Futura”粗体小写字母相结合。

系统的主要特点在于电梯旁边的超大尺寸数字的使用。这些从地板延伸直至天花板的数字去掉橙黄色的遮盖，显现出原始的粗糙混凝土墙。

除标识系统外，卡特利吉・列文还制作了一份简单的折叠风琴式地图，作为访客在中心的导航，在薄纸的背面使用镜像。

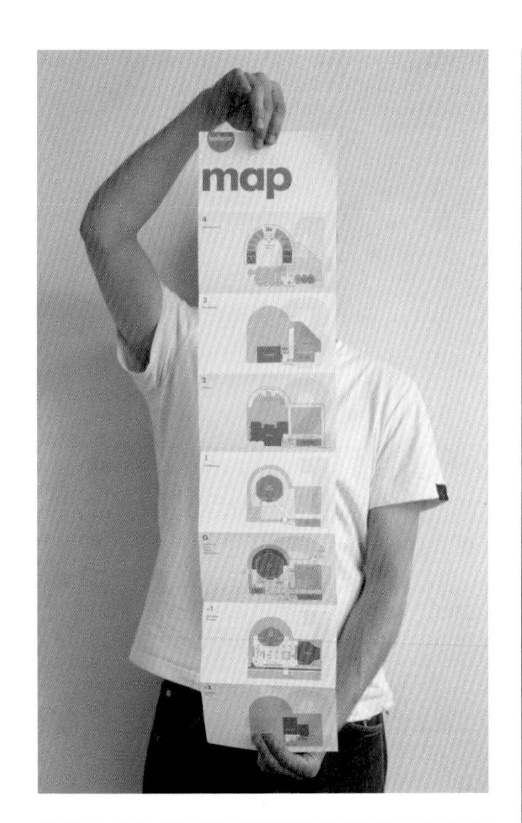

▸ 01

设计者：Studio Myerscough / Cartlidge Levene
作品名：*Barbican Arts Centre, London*
导示系统

▸ 02

设计者：Cartlidge Levene
作品名：*Map*
楼层平面图

CITY
OF
LONDON
cinemas 2 & 3
conference rooms
3 art gallery
library
restaurants
hall
curve
tickets
theatre
cinema 1
pit
3

07.2.2
超越边界

为宣布即将在维也纳应用艺术大学建筑学院举办的第四届城市原型设计会议而制作了该宣传资料单。A5的传单折叠成A2海报，显著的数字“4”填满了整个传单页面。

The Institute of Architecture at the University of Applied Arts Vienna and Architecture Live 4 pr

urban
prototyp
conferer
curated by Ilka & Andreas
29–30 June 2007, MAK Le
University of Applied Arts
Institute of Architecture

Speakers include: Lacaton & Vassal Architectes, France | Bjarke Ingels,
Mathias Müller, Daniel Niggli, EM2N, Switzerland | Alfredo Brillembourg
Klumpner, Urban Think Tank, Venezuela | Hitoshi Abe, Atelier Hitoshi Abe,
Japan/UCLA, L.A. USA | Minsuk Cho, Mass Studies, Seoul, Korea | Alexa
MIT, USA | François Roche/R&Sie(n), Paris | Mark Wigley, Columbia Uni

www.dieangewandte.at/archurban

PROGRAM

di:'angewandte SpringerWienNewYork

▸ 01

设计者：Paulus M. Dreibholz
作品名：*Urban Prototyping Conference*
A2折叠宣传单

RAUM NETZ Vol. 2:
urban
prototyping
conference
curated by Ilka & Andreas Ruby
29–30 June 2007, MAK Lecture Hall
University of Applied Arts Vienna
Institute of Architecture
Speakers include:
die:'Angewandte
SpringerWienNewYork
The Institute of Architecture at the University of Applied Arts Vienna
urban
protot

07.2.2
超越边界

国际设计奖项图书《Focus Open 2011》（参见第16页）在开篇就表现出清晰的比例感。排版拍摄于大型工作室背景之下，看起来像是由一个四处走动为标题添加字体板块的人体构成。

“SEA设计”制作了一份大型的丝网印刷海报，用于宣布在贝尔法斯特大学举行的对其设计作品进行展示的展览和讲座。通过高度剪裁的标题及对比鲜明的小字体来夸大尺度感。

▸ 02

设计者：Stapelberg & Fritz
作品名：*Focus Open 2011*
图书设计

▸ 03

设计者：SEA Design
作品名：*Belfast University*
展览/讲座海报

Modul

T44T43N0

07.2.2
超越边界

“468工作室”是都柏林一处艺术家培训项目所在地。关于这个工作室报道的小传单/包书纸被折起以遵循每个板块（4单位/6单位/8单位）的相对比例。经裁剪成片的数字为封面创造出一种引人注目的戏剧性排版方案。

04
设计者：Atelier David Smith
作品名：*Studio 468*
折叠的报道小传单

studio 468
A working space in Rialto
468

07.2.3
全篇幅

“大十联盟”是美国最古老、最大型的一级体育协会。“五角设计”设计出具有当代校园字体特色的新标识，在“BIG”这一单词中嵌入数字10，这使得球迷们可以在一个简单的词中既可见到“BIG”又能见到“10”。

《70毫米——比生活更大》是一次回顾展，在柏林电影电视博物馆举行，展示的是用70毫米超宽电影胶片拍摄的影片。“五角设计”通过印制超出海报边缘的极大数字“70”来强调版式的大尺度。

01

▶ 01

设计者：Pentagram
作品名：*B1G*
“大十联盟”品牌重塑

▶ 02

设计者：Pentagram
作品名：*70mm — Bigger than Life*
海报

RETROSPEKTIVE
5. BIS 15. 2. 2009
International, Karl-Marx-Allee
Cinestar 8, Potsdamer Straße
Vorträge und Gespräche
www.deutsche-kinemathek.de
59.
Internationale
Filmfestspiele
Berlin
DEUTSCHE
KINEMATHEK
MUSEUM
FÜR FILM UND
FERNSEHEN
70 mm – Bigger than Life
L'ORÉAL
PARIS

02

07.2.4
三维数字

“麦耶斯考夫工作室”（Studio Myerscough）为东伦敦肖尔迪奇一处复杂的创意工作室“茶建筑”设计了引导标识及指示系统，这些都基于经常出现在传统茶箱上的钢印衬线字体设计而来。

“麦耶斯考夫工作室”也为设计杂志《ICON》的封面设计了手工版式雕刻《2020》。白色刻字上的白色似乎创造出位图字体的一种模拟版本，这些依赖于光与影的可识别性。《数字2》对这种三维点阵图字体主题进行了进一步探索，对不同色彩的木块加以布置以形成数字“2”。

01

02

▸ 01
▸ 02

设计者：Studio Myerscough
作品名：*Tea Building*
引导标识及指示系统

▸ 03

设计者：Studio Myerscough
作品名：*2020*
《ICON》杂志封面

▸ 04

设计者：Studio Myerscough
作品名：*Number 2*
三维字母样品

03

04

07.2.5
数字挡字书板

“光州双年展”是亚洲首个当代艺术双年展，于1995年在韩国光州正式推出。“Base Design”对2008年的双年展标识进行了设计。《年度报告：一年中的展览》重点突出2007年1月至2008年9月所有双年展中的艺术作品展。设计者将“07”和“08”形象地设计成将任何类型或尺寸集合在一起的挡书板。由此，展览的内容便成为了标识本身的一部分。

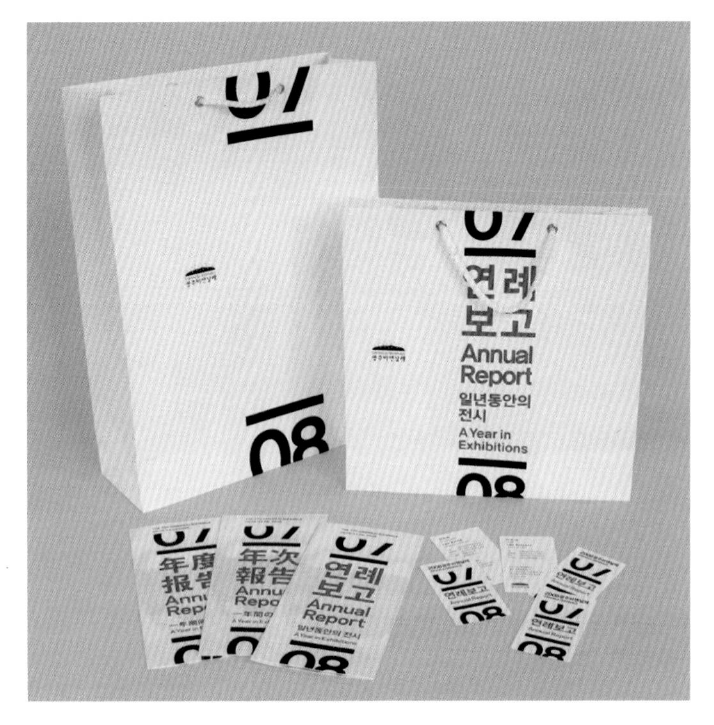

▶ 01

设计者：Base Design

作品名：*Gwangju Biennale*

引导标识、海报、图书及各式各样的印刷制品

연례보고

Annual Report

일년동안의 전시

A Year in Exhibitions

길 위에서
On the Road

제안
Position Papers

끼워넣기
Insertions

프로그램 Programs

글로벌 인스티튜트
Global Institute

국제학술회의
Plenary Sessions

08

2008광주비엔날레
GWANGJU BIENNALE
2008.9.5–11.9

중외공원 비엔날레관, 광주시립미술관, 의재미술관, 광주극장, 대인시장
BIENNALE HALL, GWANGJU MUSEUM OF ART, UIJAE MUSEUM OF KOREAN ART, CINEMA GWANGJU, DAEIN TRADITIONAL MARKET
주최: 재단법인 광주비엔날레, 광주광역시

재단법인 광주비엔날레, 광주광역시 북구 비엔날레 2길 211
GWANGJU BIENNALE FOUNDATION, 211 BIENNALE 2(I)-GIL, BUK-GU, GWAGNJU 500-070, KOREA
TEL +82 (0)62-608-4114 / WWW.GB.OR.KR

08.
减少

08.1 缩减
08.1.1 线条与形式 292
08.1.2 简化与削减 294
08.1.3 图形形状 296
08.1.4 裁剪与旋转 298

08.2 简化
08.2.1 剪裁 300
08.2.2 切片 302
08.2.3 少即是多 304
08.2.4 直线和曲线 306

08.1
缩减

08.1.1
线条与形式

“五角设计”每年都发行一本小型图书“假日”来表示对其朋友、客户及同事的问候。工作室内的设计团队轮流对图书进行研究和设计，图书通常包含某种游戏或活动。麦克·格里克及其团队设计了作品《大量的数字》。数字是一个特别的适时性主题，鉴于他们在近期世界性事件及联络方式上的重要性，现在比以往更加如此。该图书的特色在于它的七个数字体系，这里显示的是从简单的计数到缅文系统。

“第八工作室”设计海报《排版和印刷的合作》用来在伦敦进行的“排版圈儿”活动中发表演讲。这种简单的设计从工作室的Logo中提炼出黑圈元素并将其变形为大号数字“8”。

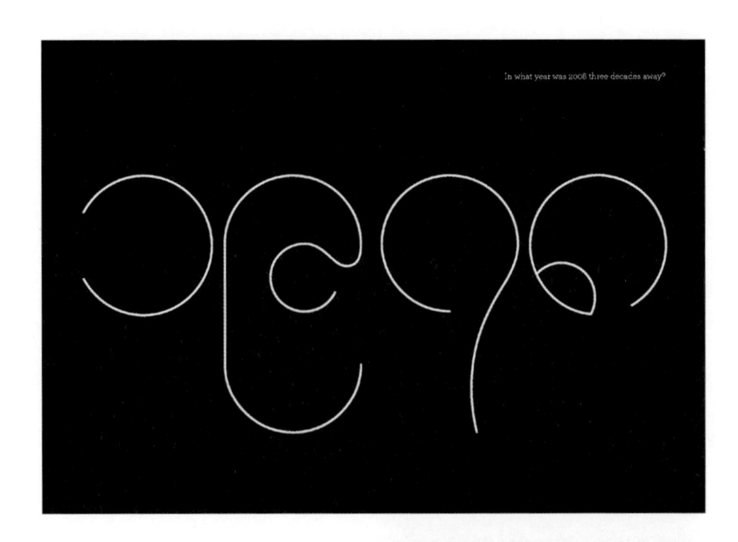

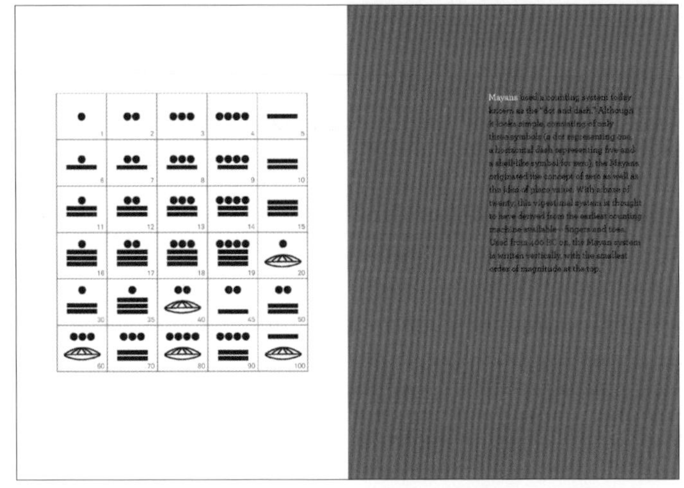

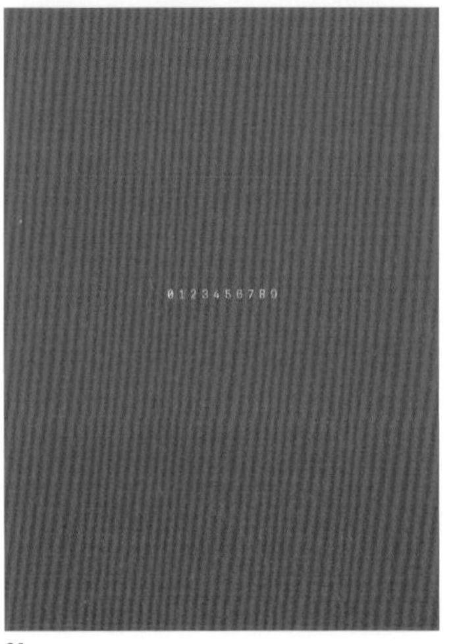

01

▸ 01

设计者：Pentagram
作品名：*A Number of Numbers*
来自世界各地的数字

▸ 02

设计者：Studio8 Design
作品名：*Collaboration in type & Print*
为“排版圈儿”活动设计的海报

Studio8 Design:
Collaborations in
Type & Print
27 May 2010
7pm
The Typographic Circle
JWT, 1 Knightsbridge Green
London SW1X 7NW
www.typocircle.co.uk
Paper by:
GF Smith
Printed on:
GF Smith Colorplan Citrine 170gsm
Printed by:
White Duck Screenprint
typ

02

08.1.2
简化与削减

威尔·孔斯设计的一款2007新年贺卡《2K+7》，2000年被写成“2K”，其后面设计有着极细的“+”号图形，最后为数字“7”。其设计的2006年新年贺卡《放松》，通过形成数字“6”的松散弹簧来进行生动的显示。

“Struktur Design”为吉米9岁生日设计的个人卡片《吉米9》对数字“9”进行切分，并将其变成字母“Y”来完成他名字的构成。

best wishes
for the
new year

2K + 7

01

▸ 01

设计者：Willi kuns
作品名：*2K+7*
2007新年贺卡

▸ 02

设计者：Willi kuns
作品名：*Unwind*
2006新年贺卡

▸ 03

设计者：Struktur Design
作品名：*Jimmy 9*
生日卡片

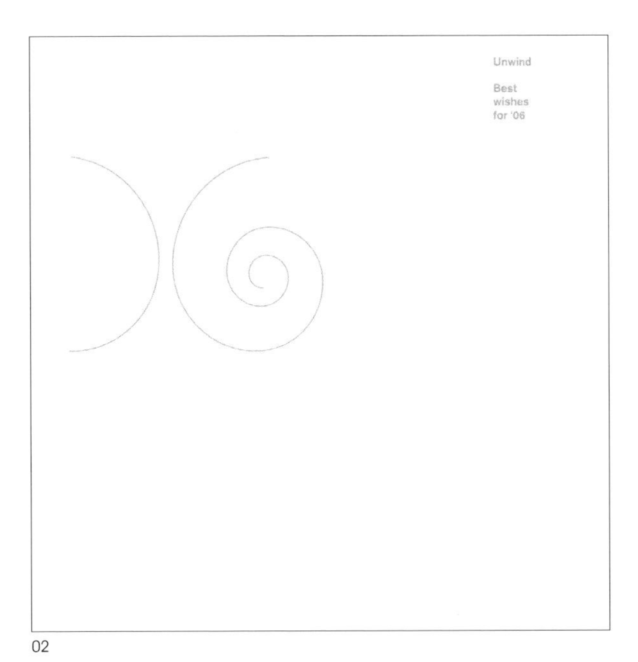
Unwind
Best
wishes
for '06

02

jimmy

happy ninth birthday jimmy. love from uncle roger / auntie sanne / tristan / minnie / monty

03

08.1.3
图形形状

《15》是莱奥纳多·索罗里（Leonardo Sonnoli）在柯林斯堡科多拉多州立大学举办的第16届“美国科罗拉多国际海报双年展”上的参赛作品。海报通过对15种彩色矩形块进行布置来形成数字“15”。

“Socket工作室”通过一场乔迁宴会来庆祝其搬入新的办公室。举办宴会的日期定于2010年10月10日——极美妙的日期图形“10.10.10”。但若建筑工程超延期，宴会只能推迟到一个不太具有如此图形性的日期进行。

01

▸ 01

设计者：Leonardo Sonnoli
作品名：*15*
16届“美国科罗拉多国际海报双年展”

▸ 02

设计者：Socket studios
作品名：*10.10.10*
搬迁新办公室

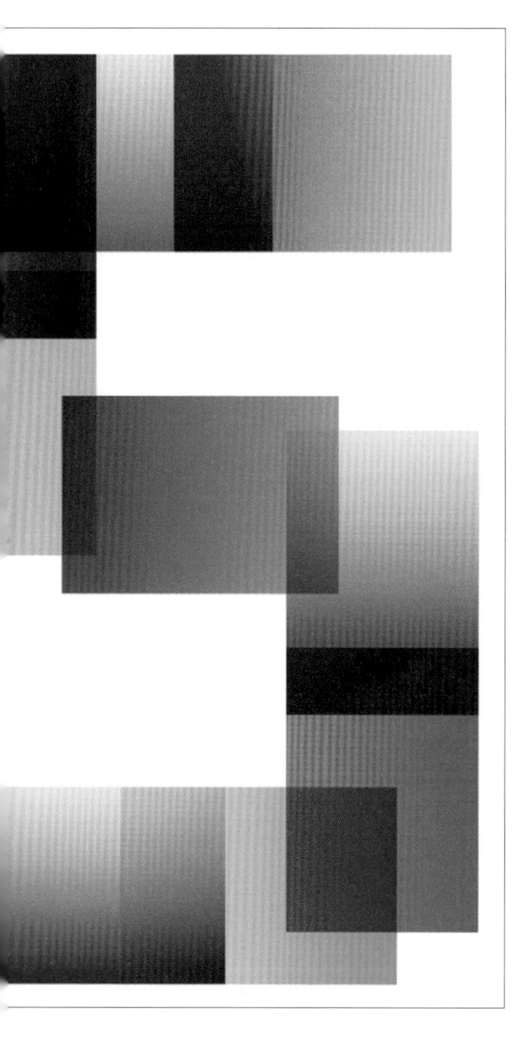

02

08.1.4
裁剪与旋转

《Guido Guidi Vitaliano Trevisan vol.I》的封面在自然裁剪上下功夫："vol.I"充满整个封面的正面和背面，当从正面看书本时仅能看到"I.I"。这为书本的标题创造出一种生动的抽象构架，书名印刷成了红色。

《Consequences at the Festivaletteratura 2010》是莱奥纳多·索罗里设计的明信片，用来宣传在意大利曼图亚进行的第四届文学节（Festivaletteratura）。这是一款漂亮的简化设计，对数字"1"进行复写、翻动及旋转来形成数字"4"。

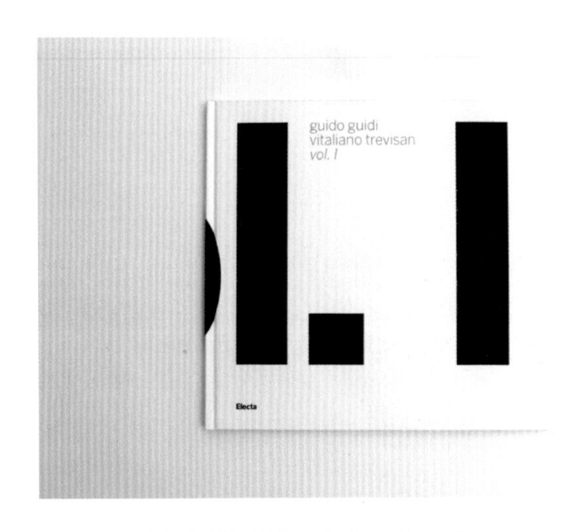

01

▸ 01

设计者：Leonardo Sonnoli
作品名：*Guido Guidi Vitaliano Trevisan vol.I*
书籍装帧设计

▸ 02

设计者：Leonardo Sonnoli
作品名：*Consequences at the Festivaletteratura 2010*
宣传明信片

08.2
简化

08.2.1
裁剪

《4NATIO第三届展览》是参访“4NATION”秋/冬男装系列设计的简单黑白邀请卡。“Newwork”设计工作室将大写字母“R”裁剪转换成许多的数字“3”来表示第三届展览会。在标题中应用这一新型的文字/数字标志，并将其作为邀请卡的主要图形。

01

设计者：Studio Newwork
作品名：*4NATION Third Exhibition*
男装系列发布会邀请卡

4NATION [four nation]
Ebina 6F 6-18-10 Jingumae
Shibuya-ku Tokyo
150-0001 Japan
T. 03 5464 9321
F. 03 5464 9331
Daisuke Tejima
M. 090 4915 0564
daisuke_tejima@gmpc-japan.com
Ayumi Kurumoto
M. 090 7646 3616
ayumi_kurumoto@gmpc-japan.com
Exhibition
2007-2008 For Men

Thi3d Exhibition

08.2.2
切片

《62》是一张62岁生日卡片，客户的出生年月被裁剪并分成片段来显示她的当前年龄。通过洋红与鲜明的黑白对比来加以突出。

《创意2007》是为在迪拜举行的地区创意峰会设计的海报（参见第110页）。活动的标题采用裁切式的字体形式，“A”和“S”形成一根镜像垂直轴，在年度的两个数字“0”中相互呼应。

01

▸ 01

设计者：Struktur Design
作品名：*62*
生日卡

▸ 02

设计者：Struktur Design
作品名：*Ideas 2007*
海报

ideas 2007
Ideas2007: Creativity Beyond Borders
The First Middle East and North Africa Creativity Congress
www.cbb2007.com
creativity
beyond
borders

02

08.2.3
少即是多

赫尔穆特·舒密特在庆祝中国成功申办2008年奥运会的宣传设计中，所有颜色都被去除，白色底板上黑色字体的应用使得设计的纯粹性得以完全表现，不附带任何干扰或令人眼花缭乱的元素。日期被重复使用，删除这一年数字的中央部分，剩下重复的曲线——奥运五环。

威利·孔茨（Willi Kunz）设计的《2002》进一步回到条状造型。对数字“2”进行水平切分，当作第一个及最后一个数字来用；同样地，对数字“0”进行垂直切分以形成中央的两个数字。设计的经济性通过“多快好省”这一信息来加以突出。

拉特利吉·列文（Cartlidge Levene）设计的圣诞及新年卡片是来自桦木胶木板的激光切割。该设计使用粗钢网衬线体，每年度的阅读方向是相反的，以此创造出一款双朝向的卡片。

2008

a journey
of a thousand miles
starts with
a single step

lao tse

congratulations
to the olympic games
beijing 2008

helmut schmid
helmut schmid design
osaka japan

2001 07 19

01

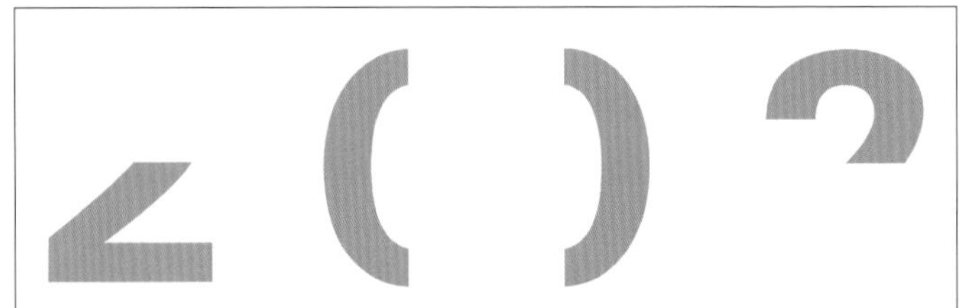

2002 do more with less

02

▸ 01

设计者：Helmut Schmid
作品名：*2008 Beijing Olympics*
海报

▸ 02

设计者：Willi Kunz
作品名：*2002*
新年贺卡

▸ 03

设计者：Cartlidge Levene
作品名：*2011－2012*
圣诞及新年卡片

03

08.2.4
直线和曲线

“Paulus M. Dreibholz”设计了一系列限量版A1丝网印刷海报，来宣传将在维也纳应用艺术大学举行的“建筑生活”活动。海报将分发给2004至2006这三年里的演讲嘉宾。

海报将数字形式简化到它们的基本元素“线”与“弧”。通过将图形拆裂成黄色的两块，使数字变得更加抽象化。

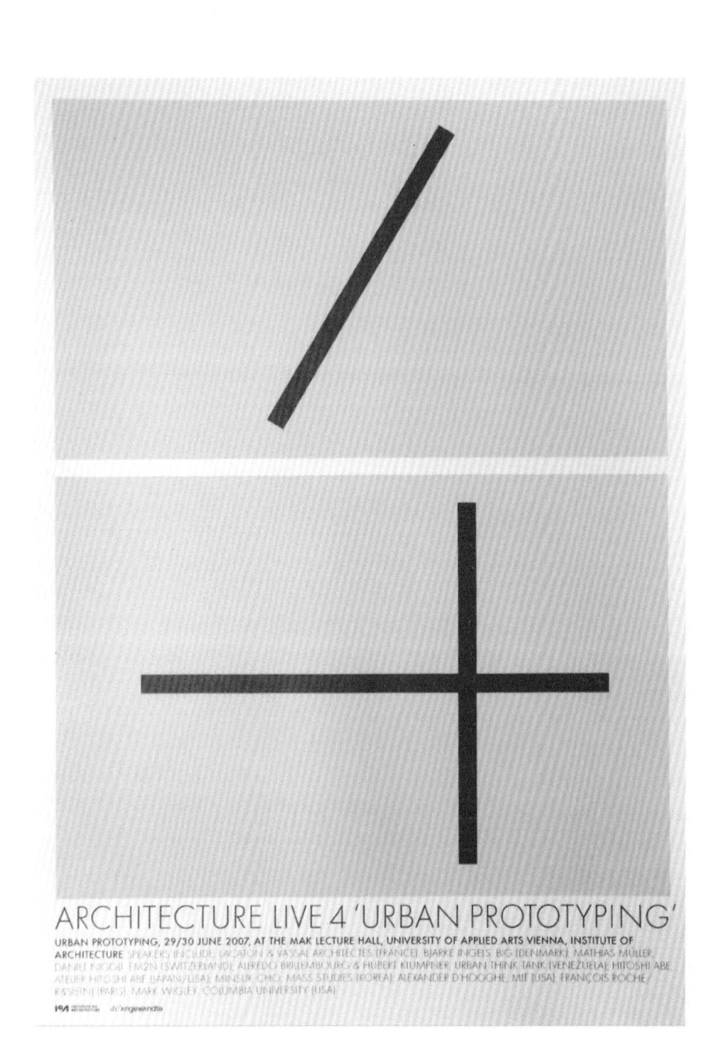

▶ 01

设计者：Paulus M. Dreibholz

作品名：*Architecture Live*

为维也纳应用艺术大学设计的海报

ARCHITECTURE LIVE 6 'ALUMNI'

ALUMNI – NETWORKS REVISITED, 14/15 MAY 2009, AT THE UNIVERSITY OF APPLIED ARTS VIENNA, INSTITUTE OF ARCHITECTURE
HITOSHI ABE, CECIL BALMOND, HERWIG BAUMGARTNER, ERIK BERNHARD, ALFREDO BRILLEMBOURG, IRMGARD FRANK, MARIE THERESE HARNONCOURT, HUBERT HERMANN, BARBARA IMHOF, JEFFREY KIPNIS, HUBERT KLUMPNER, SYLVIA LAVIN, BART LOOTSMA, CAREN ORHALING, MAX RIEDER, WOLFGANG TSCHAPELLER, THOMAS VIZKE, MICHAEL WALRAFF, CARMEN WIEDERIN, SUSANNE ZOTTL

di:'angewandte

08.2.4
直线和曲线

在为萨拉设计的35岁生日海报《35萨拉》中，数字“5”变成了字母“S”。为纪念68岁生日，设计师使用了四个圆圈。通过对一个圆圈进行四分之一的切割并将切割部分上移，便形成了数字“6”；数字“8”为简单的两个圆圈，一个位于另一个之上。

02

▸ 02

设计者：Struktur Design
作品名：*35arah*
生日海报

▸ 03

设计者：Struktur Design
作品名：*68*
生日海报

happy birthday em / grandma
love from sanne / roger / tristan / minnie / monty
90

03

1997
001 01 w 01 jan 01 CH|D|DK|E|F|I|UK|USA
002 02 t
003 03 f
004 04 s
005 05 s
006 06 m 02
007 07 t
008 08 w
009 09 t
010 10 f
011 11 s
012 12 s
013 13 m 03
014 14 t
015 15 w
016 16 t
017 17 f
018 18 s
019 19 s
020 20 m 04
021 21 t
022 22 w
023 23 t
024 24 f
025 25 s
026 26 s
027 27 m
D|E|I
USA
glædelig jul
merry christmas
23.59.59
359 24 t
360 25 w
361 26 t
366 31 t

09.
附录

09.1
规则+细节

09.1.1
非线性数字

传统正文文本使用非线性数字或旧式数字。图01说明某个设为线性数字的示范文本，即使数字采用的是大写体，它们似乎也有要跳出文本的倾向。图02显示的是设为非线性数字（旧式数字）的相同文本。在该例子中，数字遵循正文文本的同一节奏且具有更和谐的效果。

使用于正文时，非线性数字的优雅细节往往会丢失，然而用于标题时则会显示其真实美。

例03显示的是非线性数字与小写字体之间的关系及其结合使用时所具有的舒适性。

Cum venim augait nullaorperat lam iustrud magna faccum nullan et dolore eugait la acilit augait at. Ud dolendre etum nim delesed dio con velit vel eratem quat nim num (fig. 13.7) iliquam, consequ iscidunt ute ming ercin eugiam, volorem ipis atie etue venit adipisit (fig. 48) veriure tion erillan ut do od ex er suscil ea feu consenissi. Quamet autpat, sequipismod magnim dipit nonse 1958 to 1958 deliqui blaortinim diamcommy nim zzriliquat vel utpat wis aliquat lutem zzrit lum delenisim in volor augait adio dolorpe rcilit vel ipit ut wisl esequat. Ugiam, quismod magna (see pp.82/83) eugait eum vero odit ute consequi ilismod tem zzrilisisi. Raestrud magna am dolor sit la commolorer si tin et, quipsum sandipi smodolo boreet, commodo (pp.184/85) lortisi volor.

01

Cum venim augait nullaorperat lam iustrud magna faccum nullan et dolore eugait la acilit augait at. Ud dolendre etum nim delesed dio con velit vel eratem quat nim num (fig. 13.7) iliquam, consequ iscidunt ute ming ercin eugiam, volorem ipis atie etue venit adipisit (fig. 48) veriure tion erillan ut do od ex er suscil ea feu consenissi. Quamet autpat, sequipismod magnim dipit nonse 1958 to 1958 deliqui blaortinim diamcommy nim zzriliquat vel utpat wis aliquat lutem zzrit lum delenisim in volor augait adio dolorpe rcilit vel ipit ut wisl esequat. Ugiam, quismod magna (see pp.82/83) eugait eum vero odit ute consequi ilismod tem zzrilisisi. Raestrud magna am dolor sit la commolorer si tin et, quipsum sandipi smodolo boreet, commodo (pp.184/85) lortisi volor.

02

▸ 01

以线性数字设计的鲍尔·博多尼（Bauer Bodoni）活字字体

▸ 02

以非线性（旧式）数字设计的鲍尔·博多尼（Bauer Bodoni）活字字体

▸ 03

博多尼古体字非线性（旧式）数字，显示基线、x坐标、字母的上半部及下伸部。

▸ 04

“shownBodoni”博多尼古体字线性（旧式）数字，显示基线、x坐标、字母的上半部及下伸部。

▸ 05

在标题中设置鲍尔·博多尼（Bauer Bodoni）活字字体非线性（旧式）数字

0123456789xXg

03

0123456789xXg

04

52013836
4789

05

09.1.2 等宽字体与表格

虽然非线性数字与正文在一起十分和谐，且线性数字非常适合在一般情况下使用，但等宽数字最适合列表数据的排列。等宽数字与线性数字的不同之处在于等宽数字间具有固定的宽度。通常来说，数字“1”所占的位置远小于其他数字所占的位置。当需要将一列数字排到栏目中去时，这是有问题的——是设计公司报表、账目、时间表或事件列表的所有人都需要面对的问题。

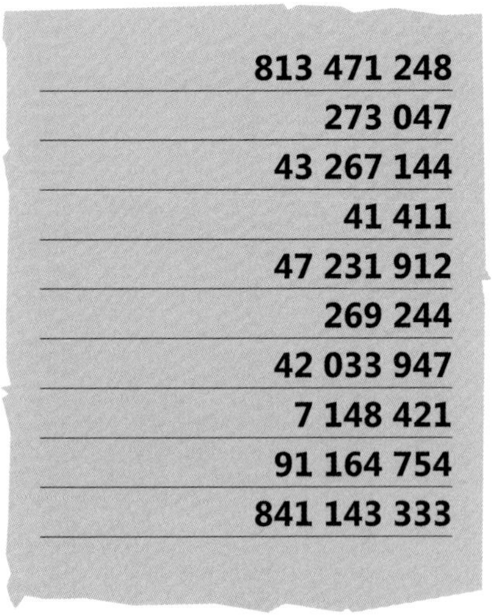

813 471 248
273 047
43 267 144
41 411
47 231 912
269 244
42 033 947
7 148 421
91 164 754
841 143 333

01

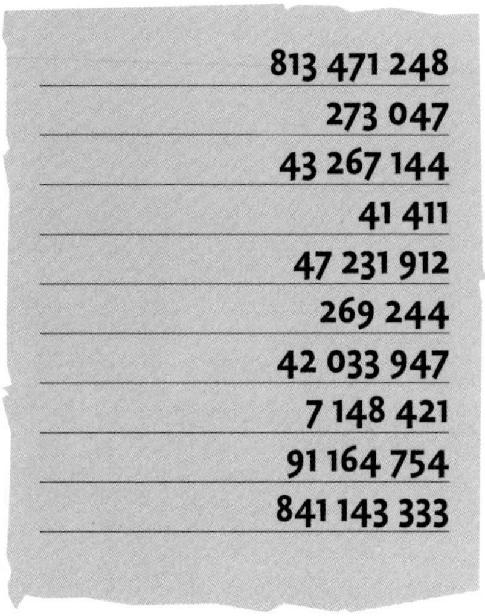

813 471 248
273 047
43 267 144
41 411
47 231 912
269 244
42 033 947
7 148 421
91 164 754
841 143 333

02

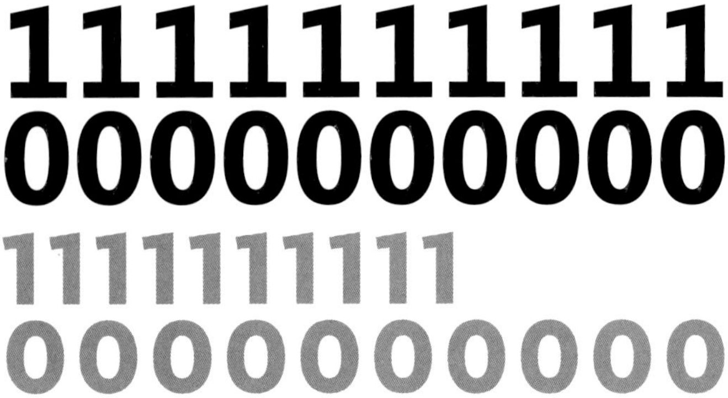

1111111111
0000000000
1111111111
0000000000

03

▸ 01
“FF Profile”不加粗体等宽字体，非常适合列表数据。

▸ 02
具有非线性数据的“FF Profile”加粗字体；
由于数字无法垂直对齐，所以很难处理列表数据。

▸ 03
设为“FF Profile”不加粗字体（上部分）和“FF Profile”加粗字体（下部分）的一系列数字“0”和数字“1”。

▸ 04
为突出字间间隔的巨大差异，对“FF Profile”加粗字体的非线性数字（这里以灰颜色出现）与“FF Profile”不加粗字体（一种等宽字体）进行黑色套印，采用固定的单位宽度。

1111111111
2222222222
3333333333
4444444444
5555555555
6666666666
7777777777
8888888888
9999999999
0000000000

04

09.2
索引

AboutTime (Chilli X) 154
A_B_Peace & Terror Etc. (The Luxury of Protest) 50, 51
Accept and Proceed
Average UK Rainfall 80
Hours of Light and Hours of Dark series 24, 25, 26
K2 Sporting Moments 2010 series 18–23
Acme Studios: *Hong Kong Handover 1997* series 120–3
Ahrend Four_Two (SEA Design) 184–5
Alie Street signage system (Commercial Art) 260–1
Amoeba (Jób) 70
Anniversary Dinner invitations (Eggers + Diaper) 250
Applied Works
Ben Saunders Polar Explorer 82
The Times iPad Edition series 68–9
April Graphics
John McAslan + Partners signage system 268–9
London Mayor's Fund 66–7
Architecture Live series (Dreibholz) 308–9
Arcoprint by Fedrigoni (Design Project) 42, 43
Artiva Design
Dopo la rivoluzione 212
45° 224, 225
Lettera22 206–7
Askul range (Stockholm Design Lab) 242, 244–5
At This Rate (Studio8) 54
Atelier Atli Hilmarsson
HönnunarMars 2009 98–9
Public Transport of Ísafjörður 94–5
Reykjavik 871+-2 226–7
Atelier David Smith
Firestation Workshops 2004 164, 165
Raw Alternatives 230, 231
650-1575: Images of a Mine 230
Studio 468 284–5
10th Anniversary Book 230
Atelier Works: *Connect Sheffield* 96–7
Audi A8 (Stapelberg & Fritz) 46, 47
Average UK Rainfall (Accept and Proceed) 80

Barbican Arts Centre signage system (Studio Myerscough/Cartlidge Levene) 278–9
Base Design
Gwangju Biennale 290–1
La Casa Encendida 114–15, 142, 146–7
Pass scientific adventures park signage system 270–1
Belfast University (SEA Design) 282, 283
Bella Sky signage system (Stockholm Design Lab/Thomas Eriksson Architects) 272, 273
Ben Saunders Polar Explorer (Applied Works/Studio8 Design) 82
Bibliothèque
Design: Inside Out 110
Happy Birthday Massimo 234
07/08 Annual Review 86
Super Contemporary 12–13
Big (Pentagram) 286
Big Bang (Jób) 70, 71
Birth Clock (Studio Tonne) 148, 149
Blanka-Helvetica 50 (Build Design) 42
Bloom, Mark: *This is my Really Useful Poster* series 110
Bloomberg Businessweek (Working Format/Commercial Type) 248–9
Bloomberg signage system (Pentagram) 262–5
Breaking Barriers Report (Commercial Art) 88, 89
Browns: *Hiscox 2010* report 84, 85
Bugs by the Numbers (Werner Design Works) 174–5
Build Design
Blanka-Helvetica 50 42
GP is 10 172–3
Phil Harrison Award 182–3
Bullseye (Lárusson) 124
Bunch Design
Consultant Group 2010 234
Eurotower 2 190, 191
555 Poster 170, 171
55DSL Tote Bag 170
1212NYC 190
Typographic Desk Calendar 176
Business turnover Berlin 2007 (Von B und C) 56

calendars (Struktur Design) 130–1
calendars (This Studio) 138–9
Cartlidge Levene
Barbican Arts Centre signage *system 278–9*
A Graphic Odyssey Poster 100–1
Lend Lease 224
2011-2012 306, 307
Chic magazine (Marnich Associates) 236
Chilli X: iPhone clock series 154
City ID: *Connect Sheffield* 96–7
Clarke, Malcolm: 30th *Anniversary of Grays Antiques* 212
ClickClock (Coffeecoding) 156
Coffeecoding: *ClickClock* 156
Collaborations in Type & Print (Studio8 Design) 294, 295
Columbia Architecture Planning Preservation (Kunz) 108, 109
Columbia University Graduate School (Kunz) 108
Commercial Art
Alie Street signage system 260–1
Breaking Barriers Report 88, 89
Commercial Type: *Bloomberg Businessweek* 248–9
Commune Inc: *TextClock* 156
Connect Sheffield (City ID, Atelier Works and Pearson Lloyd) 96–7
Consequences at the Festivaletteratura 2010 (Sonnoli) 300, 301
Consultant Group 2010 (Bunch Design) 234

Daimler Chronik (L2M3 Kommunikationsdesign) 30–1
Data Visualization (Von B und C) 62, 63
Delahunty, Jason
The Inauguration of President Barack Obama 64
Unix Time Stamp series 64–5
Design: Inside Out (Bibliothèque) 110
Design Center Stuttgart series (Stapelberg & Fritz) 104, 105
Design Project
Arcoprint by Fedrigoni 42, 43
Fedrigoni calendar 126–7
Design Scene Berlin (Von B und C) 56
DesignMarch 2009 (Atelier Atli Hilmarsson) 98–9
Development of the Berlin Cluster (Von B und C) 56, 57
Distributel 2 (Stapelberg & Fritz) 116–17
Dopo la rivoluzione (Artiva Design) 212
Doyle Partners
15 One Two 186, 187
30 Years Anniversary Cover 188
Dreibholz, Paulus M.
Architecture Live series 308–9
Sliver Lecture Series 2009 102
Urban Prototyping Conference 280–1
Dublin Bus (Image Now) 92–3

E&A Magazine (Stapelberg & Fritz) 86, 87
ECV identity (Ich & Kar) 256–7
Eggers + Diaper
Anniversary Dinner invitations 250
Max Frisch 1911-2011 140–1
10 + 5 = Gott 202–3
Eggertsson, Siggi: *Public Transport of Ísafjörður* 94
87 (Ellery) 196–7
88 Morningside (Pentagram) 198, 199
1882 (Pentagram) 234, 235
Elephant magazine (Studio8 Design) 238, 239
Ellery, Jonathan: *87* 196–7
Encounter (Stapelberg & Fritz) 46
Eurotower 2 (Bunch Design) 190, 191
Everyone Ever in the World (The Luxury of Protest) 72–3
Exquisite Clock (Fabrica/Joao Wilbert) 188, 189

F-Clock (Winfield & Co.) 154, 155
Fabrica: *Exquisite Clock* 188, 189
Fanelli, Sara: *If You Could Collaborate* series 178–81
Farrow Design
Pet Shop Boys – Disco Four 192–3
SCP clock series 158, 159
Fedrigoni calendar (Design Project) 126–7
Fedrigoni calendar (Studio8 Design) 124, 125
FF X-TRA (Sonnoli) 176, 177
15 (Sonnoli) 298
15 One Two (Doyle Partners) 186, 187
Fifth anniversary Russian disco (gggrafik) 254, 255
52 Weeks (Struktur Design) 134, 135
Finetime (Farrow Design) 158, 159
Firestation Workshops 2004 (Atelier David Smith) 164, 165
555 Poster (Bunch Design) 170, 171
55DSL Tote Bag (Bunch Design) 170
Focus Open 2011 (Stapelberg & Fritz) 164, 282
Form
Media Trust report 84
187 Lockdown series 194–5
Formula One (Studio Myerscough) 10–11
45° (Artiva Design) 224, 225
4527 (Schmid) 218, 219
4NATION Third Exhibition (Studio Newwork) 302–3
Framtida Typeface (Grandpeople) 198
Frozen Dreams (Struktur Design) 32
Fundació Palau fifth anniversary (Marnich Associates) 218–19

gggrafik
Fifth anniversary Russian disco 254, 255

24 254
GP is 10 (Build Design) 172–3
GQ Italia (Sonnoli) 254
Grandpeople: *Framtida Typeface* 198
Graphic Design Essentials (Struktur Design) 246–7
A Graphic Odyssey Poster (Cartlidge Levene) 100–1
Great Moments of a Patron (L2M3 Kommunikationsdesign) 16–17
Grootens, Joost
I swear I use no art at all 28–9
Vinex Atlas 78–9
Guido Guidi Vitaliano Trevisan Vol. I (Sonnoli) 300
Gwangju Biennale (Base Design) 290–1

Happy Birthday Massimo (Bibliothèque) 234
Hardie, George: *If You Could Collaborate* series 178–81
Heys, Simon: *WordClock* 156, 157
Hinex series (Schmid) 242, 243
Hiscox 2010 report (Browns) 84, 85
Hong Kong Handover 1997 series (Acme Studios) 120–3
HönnunarMars (Atelier Atli Hilmarsson) 98–9
Hours (Struktur Design) 144
Hours of Light and Hours of Dark series (Accept and Proceed) 24, 25, 26
How Long are 24 Hours Anyway? (Studio Tonne) 152–3
Hughes, Rian: *TimeDevice 01* 154

I swear I use no art at all (Grootens) 28–9
Ich & Kar
ECV identity 256–7
Kernia Bar 112, 113
One Strip 104
Icon Watches (SEA Design) 158
Ideas 2007 (Struktur Design) 112, 304, 305
If You Collaborate series (Sonnoli, Fanelli and Hardie) 178–81
Image Now
Dublin Bus 92–3
Muhammad Ali v George Foreman 26, 27
The Inauguration of President Barack Obama (Delahunty) 64
ION (Jób) 150
iPhone clock series (Chilli X) 154
iPhone clock series (Winfield & Co.) 154, 155
ISIA Posters (Sonnoli) 220–1

Jimmy 9 (Struktur Design) 296, 297
Jób, Ondrej
Amoeba 70
Big Bang 70, 71
ION 150
John McAslan + Partners signage system (April Graphics) 268–9

K2 (SEA Design) 214–15
K2 Sporting Moments 2010 series (Accept and Proceed) 18–23
Kernia Bar (Ich & Kar) 112, 113
Kingsbury Press cards (Thompson Brand Partners) 216, 217
Kreissparkasse, Ludwigsburg signage system (L2M3 Kommunikationsdesign) 276–7
Kunz, Willi
Columbia Architecture Planning Preservation 108, 109
Columbia University Graduate School 108
2002 306, 307
2K + 7 296
Unwind 296, 297

La Casa Encendida (Base Design) 114–15, 142, 146–7
Lárusson, Hörður
Bullseye 124
3Ö 232
Work Calendar 82
Work Year 2007 82, 83
Lend Lease (Cartlidge Levene) 224
Lettera22 (Artiva Design) 206–7
L2M3 Kommunikationsdesign
Daimler Chronik 30–1
Great Moments of a Patron 16–17
Kreissparkasse, Ludwigsburg signage system 276–7
Tübingen District Council signage system 80, 81
WerkStadt Dialog 14–15
London College of Communications signage system (Studio Myerscough) 274–5
London Mayor's Fund (April Graphics) 66–7
The Luxury of Protest
A_B_Peace & Terror Etc. 50, 51
Everyone Ever in the World 72–3
Maths Dreamed Universe 52–3
Real Magick in Theory and Practise 48
2Dots 50
The Universal Declaration series 74–5

Magnusson Fine Wine signage system (Stockholm Design Lab/Thomas Eriksson Architects) 272
Malcolm Webb (Studio Tonne) 150, 151
Marnich Associates
Chic magazine 236
Fundació Palau fifth anniversary 218–19
Mash Creative
Seconds, Minutes, Hours series 128–9
This is my Really Useful Poster series 110, 111
Massin Chapter Openers (Struktur Design) 236, 237
Master Selector (SEA Design) 228–9
Maths Dreamed Universe (The Luxury of Protest) 52–3
Max Frisch 1911-2011 (Eggers + Diaper) 140–1
Media Trust report (Form) 84
Metropolitan Wharf signage system (SEA Design) 266–7
Minimum series (Struktur Design) 106–7
Minnie's 8th Birthday (Struktur Design) 170
Minutes (Struktur Design) 132–3
Muhammad Ali v George Foreman (Image Now) 26, 27

Naturalis (SEA Design) 210–11
The New Passat Driving Experience (Stapelberg & Fritz) 42
Nightime (Farrow Design) 158, 159
NightTime (Chilli X) 154
Notime (Farrow Design) 158, 159
The 00s Issue New York magazine (Studio8) 48, 49
Number 2 (Studio Myerscough) 288, 289
A Number of Numbers (Pentagram) 294
The Numbers Series (Pentagram) 166–9

07/08 Annual Review (Bibliothèque) 86
150 (Sonnoli) 238
187 Lockdown series (Form) 194–5
One Million (Think Studio, NYC) 36–7
One Strip (Ich & Kar) 104
1212NYC (Bunch Design) 190
Otsuka Pharmaceutical reports (Schmid) 88

Pass scientific adventures park signage system (Base Design) 270–1
Pearson Lloyd: *Connect Sheffield* 96–7
Pentagram
Big 286
Bloomberg signage system 262–5
88 Morningside 198, 199
1882 234, 235
A Number of Numbers 294
The Numbers Series 166–9
Second Chicago Poster Biennial 2010 250, 251
70 mm – Bigger than Life 286, 287
3 at Saks Fifth Avenue 252–3
26 186
2008 Typographic Calendar 142
Pet Shop Boys – Disco Four (Farrow Design) 192–3
Phil Harrison Award (Build Design) 182–3
Prill Vieceli Cremers: *Stadion Letzigrund Zürich* 240–1
Public Transport of Ísafjörður (Atelier Atli Hilmarsson) 94–5

Questions (Von B und C) 76–7

Raw Alternatives (Atelier David Smith) 230, 231
Real Magick in Theory and Practise (The Luxury of Protest) 48
Remake Design
Student Handbook and Calendar 142, 143
Teague 232, 233
Volunteer Lawyers for the Arts 216, 232, 233
Reykjavik 871+-2 (Atelier Atli Hilmarsson) 226–7

Sagmeister Inc: *Self-Confidence Produces Fine Results* 38–9
Sammlung Philip Metz (Stapelberg & Fritz) 208–9
Schmid, Helmut
4527 218, 219
2008 Beijing Olympics 306
Hinex series 242, 243
Otsuka Pharmaceutical reports 88
SCP clock series (Farrow Design) 158, 159
SEA Design
Ahrend Four_Two 184–5
Belfast University 282, 283
Icon Watches 158
K2 214–15
Master Selector 228–9
Metropolitan Wharf signage system 266–7
Naturalis 210–11
20 Stationery 144, 145
Second Chicago Poster Biennial 2010 (Pentagram) 250, 251
Seconds, Minutes, Hours series (Mash Creative) 128–9
Self-Confidence Produces Fine Results (Sagmeister Inc.) 38–9
70 mm – Bigger than Life (Pentagram) 286, 287
The Sites of Ancient Greece (Struktur Design) 32, 33
650-1575: Images of a Mine (Atelier David Smith) 230
62 (Struktur Design) 304
68 (Struktur Design) 310, 311
Sliver Lecture Series 2009 (Dreibholz) 102
Social Networking Websites (Von B und C) 34–5
Socket Studios: *10.10.10* 298, 299
Sólveigarson, Stefán Pétur: *Public Transport of Ísafjörður* 94
Sonnoli, Leonardo
Consequences at the Festivaletteratura 2010 300, 301
FF X-TRA 176, 177
15 298
GQ Italia 254
Guido Guidi Vitaliano Trevisan Vol. I 300
If You Could Collaborate series 178–81
ISIA Posters 220–1
150 238
20th Festival International de l' 'Affiche et du Graphisme 212, 213
25 238
ST typefaces (Stapelberg & Fritz) 200–1
Stadion Letzigrund Zürich (Prill Vieceli Cremers) 240–1
Stapelberg & Fritz
Audi A8 46, 47
Design Center Stuttgart series 104, 105
Distributel 2 116–17
E&A Magazine 86, 87
Encounter 46
Focus Open 2011 164, 282
The New Passat Driving Experience 42
Sammlung Philip Metz 208–9
ST typefaces 200–1
Stockholm Design Lab
Askul range 242, 244–5
Bella Sky signage system 272, 273
Magnusson Fine Wine signage system 272
Struktur Design
calendars 130–1
52 Weeks 134, 135
Frozen Dreams 32
Graphic Design Essentials 246–7
Hours 144
Ideas 2007 112, 304, 305
Jimmy 9 296, 297
Massin Chapter Openers 236, 237
Minimum series 106–7
Minnie's 8th Birthday 170
Minutes 132–3
The Sites of Ancient Greece 32, 33
62 304
68 310, 311
Struktured Clock 160
31 Days 134

09.2
索引

35arah 310
365 series 136–7
Twenty Four Hour Clock 161
Units 222–3
Unleashed 32
Struktured Clock (Struktur Design) 160
Student Handbook and Calendar (Remake Design) 142, 143
Studio 468 (Atelier David Smith) 284–5
Studio8 Design
At This Rate 54
Ben Saunders Polar Explorer 82
Collaborations in Type & Print 294, 295
Elephant magazine 238, 239
Fedrigoni calendar 124, 125
The 00s Issue New York magazine 48, 49
2060 54, 55
Studio Myerscough
Barbican Arts Centre signage system 278
Formula One 10–11
London College of Communications signage system 274–5
Number 2 288, 289
Tea Building signage system 288
2020 288, 289
Studio Newwork: *4NATION Third Exhibition* 302–3
Studio Tonne
Birth Clock 148, 149
How Long are 24 Hours Anyway? 152–3
Malcolm Webb 150, 151
Time:Tone 150
24 148
Super Contemporary (Bibliothèque) 12–13

Tea Building signage system (Studio Myerscough) 288
Teague (Remake Design) 232, 233
10.10.10 (Socket Studios) 298, 299
10 + 5 = Gott (Eggers + Diaper) 202–3
10th Anniversary Book (Atelier David Smith) 230
TextClock (Commune Inc.) 156
Think Studio, NYC: *One Million 36–7*
30 Years Anniversary Cover (Doyle Partners) 188
3Ö (Lárusson) 232
30th Anniversary of Grays Antiques (Clarke) 212
31 Days (Struktur Design) 134
35arah (Struktur Design) 310
This is my Really Useful Poster series (Mash Creative) 110, 111
This Studio: calendars 138–9
Thomas Eriksson Architects
Bella Sky signage system 272, 273
Magnusson Fine Wine signage system 272
Thompson Brand Partners: Kingsbury Press cards 216, 217
3 at Saks Fifth Avenue (Pentagram) 252–3
365 series (Struktur Design) 136–7
TimeDevice 01 (Chilli X and Rian Hughes) 154
The Times iPad Edition series (Applied Works) 68–9
Time:Tone (Studio Tonne) 150
Tübingen District Council signage system (L2M3 Kommunikationsdesign) 80, 81
24 (gggrafik) 254
24 (Studio Tonne) 148
25 (Sonnoli) 238
26 (Pentagram) 186
24/7 The Alibi (Von B und C) 102, 103
Twenty Four Hour Clock (Struktur Design) 161
20 Stationery (SEA Design) 144, 145
2002 (Kunz) 306, 307
2008 Beijing Olympics (Schmid) 306
2008 Typographic Calendar (Pentagram) 142
2011-2012 (Cartlidge Levene) 306, 307
2020 (Studio Myerscough) 288, 289
2060 (Studio8) 54, 55
20th Festival International de l'Affiche et du Graphisme (Sonnoli) 212, 213
2Dots (The Luxury of Protest) 50
2K + 7 (Kunz) 296
Typographic Desk Calendar (Bunch Design) 176

U-Clock (Winfield & Co.) 154, 155
Units (Struktur Design) 222–3
The Universal Declaration series (The Luxury of Protest) 74–5
Unix Time Stamp series (Delahunty) 64–5
Unleashed (Struktur Design) 32
Unwind (Kunz) 296, 297
Urban Prototyping Conference (Dreibholz) 280–1

Vinex Atlas (Grootens) 78–9
Visual Atlas of Everyday Life at the Hospital (Von B und C) 58–9
Visualisation Research (Von Ompteda) 40–1
Volunteer Lawyers for the Arts (Remake Design) 216, 232, 233
Von B und C
Business turnover Berlin 2007 56
Data Visualization 62, 63
Design Scene Berlin 56
Development of the Berlin Cluster 56, 57
Questions 76–7
Social Networking Websites 34–5
24/7 The Alibi 102, 103
Visual Atlas of Everyday Life at the Hospital 58–9
Women's Phone study 60–1, 62
Von Ompteda, Karin: *Visualisation Research* 40–1

WDW Number Glasses (Werner Design Works) 176
WerkStadt Dialog (L2M3 Kommunikationsdesign) 14–15
Werner Design Works
Bugs by the Numbers 174–5
WDW Number Glasses 176
Wilbert, Joao: *Exquisite Clock* 188, 189
Winfield & Co: iPhone clock series 154, 155
Women's Phone study (Von B und C) 60–1, 62
WordClock (Heys) 156, 157
Work Calendar (Lárusson) 82
Work Year 2007 (Lárusson) 82, 83
Working Format: *Bloomberg Businessweek* 248–9

09.3
致谢

I would like to extend my deep thanks to all those who have helped in creating this book, whether by kindly submitting or for help and advice. The most interesting part of this process is always discovering new designers around the world who have a shared passion.

A special thank you goes to Jo, Laurence and Donald at Laurence King Publishing without who's quiet patience and understanding this book would not quite be finished yet. Finally a big thank you and much love to Sanne, Tristan, Minnie and Monty for their continued support and encouragement.

版权声明

版贸核渝字（2014）第68号